Advanced Building Simulation

T0186366

Advanced Building Simulation

Edited by
Ali M. Malkawi and Godfried Augenbroe

Spon Press
Taylor & Francis Group

NEW YORK AND LONDON

First published 2003
by Spon Press
270 Madison Avenue, New York, NY 10016

Simultaneously published in the UK
by Spon Press
2 Park Square, Milton Park, Abingdon, Oxfordshire OX14 4RN

Spon Press is an imprint of the Taylor & Francis Group

© 2004 Spon Press

Typeset in Sabon by
Newgen Imaging Systems (P) Ltd, Chennai, India

British Library Cataloguing in Publication Data
A catalogue record for this book is available from the British Library

Library of Congress Cataloging in Publication Data
 Advanced building simulation / edited by Ali M. Malkawi and Godfried
 Augenbroe.
   p. cm.
 Includes bibliographical references and index.
   1. Buildings—Performance—Computer simulation. 2. Buildings—
 Environmental engineering—Data processing. 3. Intelligent buildings.
 4. Architectural design—Data processing. 5. Virtual reality. I. Malkawi, Ali M.
 II. Augenbroe, Godfried, 1948–

 TH453.A33 2004
 690′.01′13–dc22   2003027472

ISBN 0–415–32122–0 (Hbk)
ISBN 0–415–32123–9 (Pbk)

Contents

Figures

Tables

Contributors

D. Michelle Addington, Harvard University, USA

Godfried Augenbroe, Georgia Tech, USA

Qingyan (Yan) Chen, PhD, Purdue University, USA

Larry Degelman, Texas A&M University, USA

Jan Hensen, Technische Universiteit Eindhoven, The Netherlands

Ardeshir Mahdavi, Vienna University of Technology, Austria

Ali M. Malkawi, University of Pennsylvania, USA

Sten de Wit, TNO Building and Construction Research Delft, The Netherlands

Zhiqiang (John) Zhai, University of Colorado, Boulder, USA

Acknowledgement

"Advanced Building Simulation" took several years to produce and our thanks go to all the contributors who showed immense dedication to the project. We are grateful for the support of the late Prof Emanuel-George Vakalo and Prof Jean Wineman. We also thank the students who helped with the production of this book, especially Ravi S. Srinivasan and Yun Kyu Yi at the University of Pennsylvania for their valuable assistance.

Prologue

Introduction and overview of field

Ali M. Malkawi and Godfried Augenbroe

Building simulation started to stand out as a separate discipline in the late 1970s. It has matured since then into a field that offers unique expertise, methods and tools for building performance evaluation. It draws its underlying theories from diverse disciplines, mainly from physics, mathematics, material science, biophysics, and behavioural and computational sciences. It builds on theories in these fields to model the physical behavior of as-designed, as-built, and as-operated facilities. At building scale, the theoretical challenges are inherent in the complex interplay of thousands of components, each with their own complex physical behavior and a multiplicity of interactions among them. The diversity of the interactions pose major modeling and computational challenges as they range from (bio)physical to human operated, from continuous to discrete, from symmetric to non-symmetric causality and from autonomous to controlled. Its ability to deal with the resulting complexity of scale and diversity of component interactions has gained building simulation a well-respected role in the prediction, assessment, and verification of building behavior. Specialized firms offer these services in any life cycle stage and to any stake holder.

Although most of the fundamental work on the computational core of building simulation was done two decades ago, building simulation is continuously evolving and maturing. Major improvements have taken place in model robustness and fidelity. Model calibration has received considerable attention and the quality of user interfaces has improved steadily. Software tools are currently diverse whereas simulation is becoming "invisible" and "complete" validation is considered an attainable goal. Discussions are no longer about software features but on the use and integration of simulation in building life cycle processes where integration is no longer seen as elusive goal; realistic part solutions are proposed and tested.

Advancements in Information Technology have accelerated the adoption of simulation tools due to the rapid decrease in hardware costs and advancements in software tool development environments. All these developments have contributed to the proliferation and recognition of simulation as a key discipline in the building design and operation process. Notwithstanding, the discipline has a relatively small membership and "simulationists" are regarded as exclusive members of a "guild". This book is a contribution to make designers, builders, and practitioners more aware of the full potential of the field.

While commercial tools are continuously responding to practitioners' needs, a research agenda is being pursued by academics to take the discipline to the next level. This agenda is driven by the need to increase effectiveness, speed, quality assurance, users' productivity, and others. Being able to realize this in dynamic design settings, on novel system concepts and with incomplete and uncertain information is the prime target of the next generation of tools. The integration of different physical domains in one comprehensive simulation environment is another. Meanwhile, different inter-action and dynamic control paradigms are emerging that may change the way build-ing simulation is incorporated in decision-making. The new developments will radically influence the way simulation is performed and its outputs evaluated. New work in visualization, dynamic control and decision-support seem to set the tone for the future which may result from recent shifts in the field. These shifts are apparent in the move from

- the simulation of phenomena to the design decision-making;
- "number crunching" to the "process of simulation";
- "tool integration" to "team deployment" and the process of collaboration;
- static computational models to flexible reconfiguration and self-organization;
- deterministic results to uncertainty analysis;
- generating simulation outputs to verification of quantified design goals, decision-support, and virtual interactions.

Although these shifts and directions are positive indicators of progress, challenges do exist. Currently, there is a disconnect between institutional (governmental and edu-cational) research development and professional software development. The severity of this disconnect varies between different countries. It is due in part to the fact that there is no unified policy development between the stakeholders that focuses and accelerates the advancement in the field. This is evident in regard to the historical divide between the architects and engineers. Despite the advancements in computa-tional developments the gap, although narrowing, is still visible. In addition, it must be recognized that the building industry is, besides being a design and manufacturing industry, also a service industry. Despite these challenges and the fact that many of the abovementioned new shifts and directions have yet to reach their full potential, they are already shaping a new future of the field.

This book provides readers with an overview of advancements in building simulation research. It provides an overall view of the advanced topics and future perspectives of the field and what it represents. The highly specialized nature of the treatment of top-ics is recognized in the international mix of chapter authors, who are leading experts in their fields.

The book begins by introducing the reader to recent advancements in building simulation and its historic setting. The chapter provides an overview of the trends in the field. It illustrates how simulation tool development is linked to the changes in the landscape of the collaborative design environments and the lessons learned from the past two decades. In addition, the chapter provides a discussion on distributed simu-lations and the role of simulation in a performance-based delivery process. After this overview and some reflections on future direction, the book takes the reader on a journey into three major areas of investigations: simulation with uncertainty, com-bined air and heat flow in whole buildings and the introduction of new paradigms for the effective use of building simulation.

Simulation is deployed in situations that are influenced by many uncertain inputs and uncertain modeling assumptions. The early chapters of the book take the reader through two topics that represent recent additions to this field of investigation: Chapter 2 concentrates on assessing the effect of model uncertainty whereas Chapter 3 concentrates on the effect of uncertain weather information. Chapter 2 illustrates that simulation accuracy is influenced by various factors that range from user interpretations and interventions to variations in simulation variables and behavioral uncertainties and validations. It discusses the main principles of uncertainty analysis and describes how uncertainty can be incorporated in building simulation through a case study. Chapter 3 discusses one particular form of uncertainty in building simulation, weather prediction. After describing some of the background to weather modeling and the Monte Carlo method, the chapter describes the two essential models for generating hourly weather data—deterministic models and stochastic models.

Chapters 4, 5, and 6 address the second topic in the book, the integration and coupling of air and heat flow. Each chapter offers a unique view on the attempt to increase overall simulation "quality". All three chapters deal with the application of Computational Fluid Dynamics (CFD) to the built environment, a sub-field of Building Simulation that is rapidly gaining acceptance. The chapters discuss variants of air flow models, their current limitations and new trends. Chapter 4 focuses on the coupling between domain-specific models and discusses computational approaches to realize efficient simulation. It discusses the advantages and disadvantages of the different levels of air flow modeling. Numerical solutions for integrating these approaches in building models are also discussed. Several case studies are illustrated to demonstrate the various approaches discussed in the chapter. Chapter 5 provides a review of widely used CFD models and reflects on their use. It points out modeling, validation and confidence challenges that CFD is facing. Chapter 6 on the other hand, provides a new perspective on the potential conflicts between CFD and building system modeling. It addresses the differences between building system modeling and phenomenological modeling. Cases from other fields and industries are used to illustrate how phenomenological studies can reveal unrecognized behaviors and potentially lead to unprecedented technological responses.

Chapters 7, 8, and 9 address new paradigms that have emerged. The three chapters each introduce a research field that may affect the deployment of simulation and address its impact on design and building services practices. Chapter 7 illustrates the concept of self-aware buildings and discusses how self-organizing buildings support simulation based control strategies. Case studies are provided to illustrate these concepts. Trends in process-driven interoperability are discussed in Chapter 8. The chapter provides an overview of the technologies utilized in the field to achieve interoperability. It illustrates existing approaches to develop integrated systems and focuses on a new initiative in design analysis integration that combines interoperability and groupware technologies.

Chapter 9 finally introduces new means of how users will interact with simulations in the future. The chapter introduces a newly defined area, which the author termed "immersive building simulation". The chapter defines the emerging area and describes its essential components, different techniques and their applications. It concludes with illustrative case studies and reflections remarks regarding the challenges and opportunities of this area. The three chapters are good examples of how simulation of the built environment will become ubiquitous, invisible, and omni-present.

Chapter 1

Trends in building simulation

Godfried Augenbroe

1.1 Introduction

The total spectrum of "building simulation" is very wide as it spans energy and mass flow, structural durability, aging, egress and even construction site simulation. This chapter, and indeed the book, will deal with building performance simulation in the narrower sense, that is, limited to the field of physical transport processes. This area of building performance simulation has its origin in early studies of energy and mass flow processes in the built environment. Meanwhile, the role of simulation tools in the design and engineering of buildings has been firmly established. The early groundwork was done in the 1960s and 1970s, mainly in the energy performance field followed by an expansion into other fields such as lighting, Heating Ventilation and Air-Conditioning (HVAC), air flow, and others. More recent additions relate to combined moisture and heat transfer, acoustics, control systems, and various combinations with urban and micro climate simulations. As tools matured, their proliferation into the consultant's offices across the world accelerated. A new set of challenges presents itself for the next decade. They relate to achieving an increased level of quality control and attaining broad integration of simulation expertise and tools in all stages of the building process.

Simulation is credited with speeding up the design process, increasing efficiency, and enabling the comparison of a broader range of design variants. Simulation provides a better understanding of the consequences of design decisions, which increases the effectiveness of the engineering design process as a whole. But the relevance of simulation in the design process is not always recognized by design teams, and if recognized, simulation tools cannot always deliver effective answers. This is particularly true in the early design stages as many early research efforts to embed "simplified" of "designer-friendly" simulation instruments in design environments have not accomplished their objectives. One of the reasons is the fact that the "designer" and the "design process" are moving targets. The Internet has played an important role in this. The ubiquitous and "instant" accessibility of domain experts and their specialized analysis tools through the Internet has de-emphasized the need to import "designer-friendly" tools into the nucleus of the design team. Instead of migrating tools to the center of the team, the opposite migration may now become the dominant trend, that is, delegating a growing number of analysis tasks to (remote) domain experts. The latter trend recognizes that the irreplaceable knowledge of domain experts and their advanced tool sets is very hard to be matched by designer-friendly variants. With this recognition, sustaining complete, coherent and expressive communications between remote simulation experts and other design team members has surfaced as the real challenge. After an overview of the maturation of the building simulation toolset in Section 1.2, we will discuss the changing team context of simulation in Section 1.3.

Simulation is also becoming increasingly relevant in other stages of a project, that is, after the design is completed. Main application opportunities for simulation are expected during the commissioning and operational facility management phases. Meanwhile, the "appearance" of simulation is changing constantly, not in the least as a result of the Internet revolution. This is exemplified by new forms of ubiquitous, remote, collaborative and pervasive simulation, enabling the discipline to become a daily instrument in the design and operation of buildings. The traditional consultancy-driven role of simulation in design analysis is also about to change. Design analysis does not exist in isolation. The whole *analysis process*, from initial design analysis request to model preparation, simulation deployment and interpretation needs to be managed in the context of a pending design, commissioning or maintenance decision. This entails that associations between decisions over the service life of a building and the deployment of building simulation must be managed and enforced explicitly across all members of the design, engineering and facility management team. A new category of web-enabled groupware is emerging for that purpose. This development may have a big impact on the simulation profession once the opportunities to embed simulation facilities in this type of groupware are fully recognized. Section 1.4 will look at the new roles that building simulation could assume over the next decade in these settings. It will also look at the developments from the perspective of performance based design, where simulation is indispensable to quantify the new "metrics" of design quality. Finally in Section 1.5, emerging research topics ranging from new forms of calibration and mold simulation to processes with embedded user behavior are briefly discussed.

1.2   The maturation of the building simulation toolset

Simulation involves the "creation" of behavioral models of a building for a given stage of its development. The development stage can range from "as-designed" to "as-built" to "as-operated". The distinction is important as correctness, depth, completeness and certainty of the available building information varies over different life cycle stages. The actual simulation involves executing a model that is deduced form the available information on a computer. The purpose of the simulation is to generate observable output states for analysis, and their mapping to suitable quantifications of "performance indicators", for example, by suitable post-processing of the outputs of the simulation runs. The post-processing typically involves some type of time and space aggregation, possibly augmented by a sensitivity or uncertainty analysis.

Models are developed by reducing real world physical entities and phenomena to an idealized form at some level of abstraction. From this abstraction, a mathematical model is constructed by applying physical conservation laws. A classic overview of modeling tasks in the building physics domain can be found in Clarke (2001). Comparing simulation to the design and execution of a virtual experiment (Figure 1.1) is not merely an academic thought experiment. The distinction between computer simulation and different means to interact with the behavior of a building can become

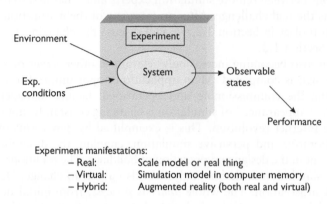

Experiment manifestations:
- Real: Scale model or real thing
- Virtual: Simulation model in computer memory
- Hybrid: Augmented reality (both real and virtual)

Figure 1.1 Simulation viewed as a (virtual) experiment.

rather blurred indeed. Interesting new interaction paradigms with simulation have emerged through combinations of real and virtual environments. This subject will resurface in later chapter in this book.

The modeling and simulation of complex systems requires the development of a hierarchy of models, or a *multimodel*, which represent the real system at differing levels of abstraction (Fishwick and Zeigler 1992). The selection of a particular modeling approach is based on a number of (possibly conflicting) criteria, including the level of detail needed, the objective of the simulation, available knowledge resources, etc. The earliest attempts to apply computer applications to the simulation of building behavior ("calculation" is the proper word for these early tries) date from the late 1960s. At that time "building simulation" codes dealt with heat flow simulation using semi-numerical approaches such as the heat transfer factor and electric network approach (both now virtually extinct). Continued maturation and functional extensions of software applications occurred through the 1970s. The resulting new generation of tools started applying approximation techniques to the partial differential equations directly, using finite difference and finite element methods (Augenbroe 1986) that had gained popularity in other engineering domains. The resulting system is a set of differential algebraic equations (DAE) derived through space-averaged treatment of the laws of thermodynamics as shown in Figure 1.2.

Since these early days, the finite element method and special hybrid variants such as finite volume methods have gained a lot of ground and a dedicated body of knowledge has come into existence for these numerical approximation techniques. Due to inertia effects, the computational kernels of most of the leading computer codes for energy simulation have not profited much from these advancements.

In the late 1970s, and continued through the 1980s, substantial programming and experimental testing efforts were invested to expand the building simulation codes into versatile, validated and user-friendly tools. Consolidation set in soon as only a handful tools were able to guarantee an adequate level of maintenance, updation and addition of desired features to a growing user base. As major software vendors continued to show little interest in the building simulation area, the developer community started to

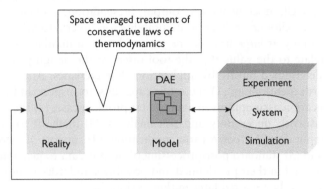

Figure 1.2 Standard approach to simulation.

combine forces in order to stop duplication of efforts. The launch of EnergyPlus (Crawley *et al.* 1999) is another more recent indication of this. Until the mid-1990s the landscape of tools was dominated by the large simulation codes that were generated with research funding, for example, DOE-2, ESP-r ad TRNSYS. As new simulation domains came along, these tools tried to expand into these domains and outgrow their traditional energy origin. However, since the late 1990s, domains other than energy are increasingly covered by specialized tools, for example, in air flow simulation, moisture and mold simulation, and others. Specialized tools do generally a better job in these specialized fields. Another new trend was the entry of commercial packages, some of which were offered as shells around the existing computation kernels mentioned earlier, and some of which were new offerings. These and all major tools are listed on (DOE 2003).

As to computational elegance, it cannot escape closer inspection that computational kernels of the energy simulation tools (still the largest and most pronounced category of building simulation tools) date back more than 15 years. Rather primitive computing principles have remained untouched as the bulk of the development resources have gone into functional extensions, user interfaces and coverage of new transport phenomena. But thanks to the fact that Moore's law (in 1965, Gordon Moore promised that silicon device densities would double every 18 months) has held over the last 25 years, current building energy simulation codes run efficiently on the latest generation of Personal Computers.

The landscape of simulation tools for the consulting building performance engineer is currently quite diverse, as a result of the hundreds of man-years that have been invested. A skilled guild of tool users has emerged through proper training and education, whereas the validation of tools has made considerable progress. As a result, the design profession appears to have acquired enough confidence in the accuracy of the tools to call on their expert use whenever needed. In spite of the growing specialization and sophistication of tools, many challenges still remain to be met though before the building performance discipline reaches the level of maturity that its vital and expanding role in design decisions demands. Many of these challenges have been on the wish list of desired tool characteristics for many years. They relate to improvements in learning curve, GUI, documentation, output presentation, animation, interactivity, modularity, extensibility, error diagnostics, usability for "intermittent" users, and others. The user community at large has also begun to identify a number of additional challenges. They relate to the value that the tool offers to the design process as a whole. This value is determined mostly by application characteristics. Among them, the following are worth mentioning: (1) the tool's capability to inspect and explicitly "validate" the application assumptions in a particular problem case; (2) the tool's capability to perform sensitivity, uncertainty and risk analyses; (3) methods to assert preconditions (on the input data) for correct tool application; (4) support of incremental simulation cycles; and (5) standard post-processing of output data to generate performance indicators quantified in their pre-defined and possibly standardized measures. Some of these challenges will be revisited later in this section.

One "development issue" not mentioned above deserves special attention. It concerns the modularity and extensibility of (large) computer codes. In the late 1980s many came to realize that the lack of modularity in current "monolithic" programs would make them increasingly hard to maintain and expand in the future. Object-oriented programming (OOP) languages such as C++ were regarded as the solution and "all it would take" was to regenerate existing codes in an OOP language. The significant advantage of this approach is the encapsulation and inheritance concepts supported by object-oriented languages. Projects attempting to use the object-oriented principles to regenerate existing programs and add new functionality were started. EKS (Tang and Clarke 1993), SPARK (Sowell and Haves 1999), and IDA (Sahlin 1996a) are the best-known efforts of that period. They started a new wave of software applications that were intended to be modular and reusable. Ten years later, only IDA (Björsell *et al.* 1999) has evolved to an industry strength application, in part due to its pragmatic approach to the object-oriented paradigm. An important outcome of these attempts are the lessons that have been learned from them, for example, that (1) reverse engineering of existing codes is hard and time consuming (hardly a surprise), (2) it is very difficult to realize the promises of OOP in real life on an object as complex as a building, and (3) the class hierarchy in an OOP application is not a "one fit all" outcome but embodies only a particular semantic view of a building. This view necessarily reflects many assumptions with respect to the building's composition structure and behavior classification. This particular developer's view may not be suitable or even acceptable to other developers, thus making the original objectives of code reusability a speculative issue. Another important lesson that was learned is that building an object-oriented simulation kernel consumes exorbitant efforts and should not be attempted as part of a domain-specific effort. Instead, generic simulation platforms, underway in efforts such as MODELICA (Elmqvist *et al.* 1999) should be adopted. An important step in the development is the creation of a building performance class hierarchy that is identical to, or can easily be mapped to a widely accepted "external" building model. The model proposed by the International Alliance for Interoperability (IAI) seems the best candidate at this time to adopt for this purpose. This is only practical however, if a "semantic nearness" between the object-oriented class hierarchy and the IAI model can be achieved. Whether the similarity in the models would also guarantee the seamless transition between design information and building performance analysis tools (a belief held by many IAI advocates, see for instance Bazjanac and Crawley (1999)) is

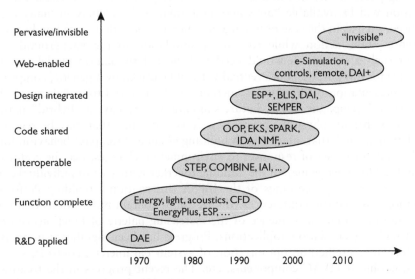

Figure 1.3 Trends in technical building performance simulation tools.

a matter that needs more study. It can be argued that such seamless transition can in general not be automated as every translation between design and analysis requires intervention of human judgment and expert modeling skills, strongly influenced by design context and analysis purpose.

In an attempt to put the observations of this section in a broad historic perspective, Figure 1.3 identifies building simulation trends between 1970 and 2010.

The foundation for building simulation as a distinct class of software applications came with the advent of first-principles-based formulation of transport phenomena in buildings, leading to DAE formulations that were amenable to standard computational methods. The next step was towards broader coverage of other aspects of technical building behavior. This movement towards function complete tools led to large software applications that are being used today by a growing user base, albeit that this user base is still composed of a relatively small expert guild. The next two major movements started in parallel in the 1990s and had similar goals in mind on different levels of granularity. Interoperability targets data sharing among (legacy) applications whereas code sharing targets reuse and inter-application exchange of program modules. Whereas the first tries to remove inefficiencies in data exchange, the latter is aiming for functionally transparent kits of parts to support the rapid building (or rather configuration) of simulation models and their rapid deployment.

Design integration adds an additional set of process coordination issues to its predecessor movements. Ongoing trials in this category approach different slices of a very complex picture. It is as yet unclear what approach may eventually gain acceptance as the best framework for integration.

The two most recent trends in Figure 1.3 have in common that they are Internet driven. The Web enables a new breed of simulation services that is offered at an increasing pace, mostly in conjunction with other project team services. Ultimately, simulation will be available "anywhere and anytime," and may in many cases go unnoticed, such as in the case of intelligent control devices that, based on known user preferences, take action while running a simulation in the background. At the moment such a simulation would need a dedicated simulation model specifically handmade for this purpose, but eventually it will be driven by a generic, complete "as-built" representation of the artifact, metaphorically referred to as "electronic building signature" (Dupagne 1991). Future simulations may have a "hybrid" nature, as they deal with both physical objects as well as occupants that may be regarded as "simulation agents" that interact with building systems. Occupant behavior, for now usually limited to a set of responses to environmental changes, may be codified in a personalized knowledge map of the building together with a set of individual comfort preferences. Recent proceedings of the IBPSA (International Building Performance Simulation Association) conferences (IBPSA 2003) contain publications that address these and other topics as evidence of the increasing palette of functions offered by current building simulation applications. Progress has been significant in areas such as performance prediction, optimization of system parameters, controls, sensitivity studies, nonlinear HVAC components, etc. The recent progress in the treatment of coupled problems is also significant, as reported in (Clarke 1999; Clarke and Hensen 2000; Mahdavi 2001).

In spite of tremendous progress in robustness and fidelity there is a set of tool functionalities that have received relatively little attention, maybe because they are very hard to realize. Some of them are discussed below.

Rapid evaluation of alternative designs   by tools that facilitate quick, accurate and complete analysis of candidate designs. This capability requires easy pre- and post-processing capabilities and translation of results in performance indicators that can easily be communicated with other members of the design team. For rapid evaluation of alternatives, tools need mechanisms for multidisciplinary analyses and offer performance-based comparison procedures that support rational design decisions.

Design as a (rational) decision-making process   enabled by tools that support decision-making under risk and uncertainty. Tools should be based on a theory for rigorous evaluation and comparison of design alternatives under uncertainty. Such a theory should be based on an ontology of unambiguously defined performance requirements and their assessments through quantifiable indicators. The underlying theory should be based on modern axioms of rationality and apply them to make decisions with respect to overall measures of building utility.

Incremental design strategies   supported by tools that recognize repeated evaluations with slight variations. These tools should respond to an explicitly defined design parameter space and offer a mechanism for trend analysis within that space, also providing "memory" between repeated evaluations, so that each step in a design refinement cycle requires only marginal efforts on the part of the tool user.

Explicit well-posedness guarantees   offered by tools that explicitly check embedded "application validity" rules and are thus able to detect when the application is being used outside its validity range.

Robust solvers for nonlinear, mixed and hybrid simulations,   going beyond the classical solving of a set of DAEs. The generation of current DAE solvers has limitations in the presence of building controls as they add a discrete time system to the overall set, leading to a mixed problem, and often to synchronization problems. Many DAE solvers fail to find the right solution in the presence of nonlinear state equations (requiring iterative solution techniques) and time critical controls (Sahlin 1996a). Hybrid problems occur when intelligent agents enter in the simulation as interacting rule-based components (Fujii Tanimoto 2003), as is the case when occupants interact with the control system on the basis of a set of behavioral rules. To cover all these cases, a robust type of multi-paradigm solvers is needed.

Two, more general, industry-wide perspectives should complete this list: certification (the end-user perspective) and code sharing (the developers perspective). Tool certification is an important aspect of QA, often perceived as enforcing the use of qualified tools and procedures. A recent PhD study (de Wit 2001) compares certification to a more general approach based on uncertainty analysis. It is argued that at best a calibration (in relation to its peers) of the combination of firm and consultant, and available tools and expertise makes sense.

Code sharing is perceived as the ultimate target of efficient collaborative code development, and object-oriented environments are considered as a pre-condition to make it happen. As introduced before, the benefits of object-oriented frameworks, for example, modularity, reusability, and extensibility are well understood. Frameworks enhance modularity by encapsulating volatile implementation details behind stable interfaces, thus localizing the impact of design and implementation changes (Schmidt 1997). These interfaces facilitate the structuring of complex systems into manageable software pieces and object-based components that can be developed and combined dynamically to build simulation applications or composite components. Coupled with diagrammatic modeling environments, they permit visual manipulation for rapid assembly or modification of simulation models with minimal effort. This holds the promise of a potentially large co-developer community as these platforms offer the capabilities to exchange whole systems or parts with other developers. Wherever co-development is practiced, it is predominantly on code level but the WWW evolution holds strong promises for functional sharing (i.e. borrowing the functions rather than the code) as well. A prime manifestation of this is distributed simulation, which is dealt with in the next section. The biggest barrier for the opportunities of shared development is the level of resources that need to be spent on the redevelopment of existing tools and the reverse engineering efforts that come with it. Unfortunately it is often regarded a safer route to expand legacy tools, but in other domains it has been shown that this approach requires more effort and produces less desirable results than a completely new design (Curlett and Felder 1995). Realistically, it must be acknowledged that most items on our list will not be realized in the very near future.

1.3   The place of simulation in the changing landscape of collaborative design teams

Building simulation has become part of the arsenal of computer applications for the design, engineering and (to some extent) operation of buildings. The primary objective of their use is to conduct a performance analysis that can inform a large number of decisions, such as design decisions, dimension parameter choices, budget allocations for maintenance, etc. How effectively this is done, is as much dependent on the quality of the tools and the technical skills of the consultant as it is on other factors. Among these other factors, the management and enforcement of the causality between certain design considerations and a requested analysis is crucial. If the interaction between design tasks and engineering analysis is incidental and unstructured the potential contribution of building simulation to achieve better buildings will not be reached. Better tuning of the coupling between design intentions and simulation deployment is needed therefore. A new category of simulation environments will emerge for precisely that purpose. The tools embedded in these environments focus on data integration and simulation interoperability but above all on rapid and timely invocation of the most adequate simulation function (rather than simulation tool) in a given design context. The two major areas of improvement for the building simulation profession can be identified as (1) tool-related, targeting advancements of tool functionality and (2) process-related, targeting functional integration of simulation tools in the design process.

1.3.1   Designer-friendly versus design-integrated tools

The development of "simplified" simulation tools for architectural designers has received a lot of attention from the research community in the past, but seems to be fading lately. Past trends were stimulated by the belief that simulation tasks should be progressively moving towards the nonspecialist, in this case the architectural designer. We argue against this and find that attempts to provide *designer-friendly* tools have been overcome by recent events, such as the WWW and the continued increase in computing power. The ubiquitous and "instant" accessibility of project partners and their advanced tools creates a stimulus to involve as many experts as desired in a design decision. These experts are expected to use the best tools of the trade and infuse their irreplaceable expertise in the communication of analysis results with other design team members. There seems to be no apparent reason to try to *des-intermediate* the domain expert. Indeed, in an era where the Internet stimulates delegation and specialization of remotely offered services, such des-intermediation appears to be counter-productive.

In every project there are distinct stages that call for different types of assessments to assist design evolution. Early conceptual design stage assessments are mostly based on expertise and experiential knowledge of consultants. To date, the current generation of simulation tools plays no significant role in this stage. If computer tools are to have an impact in this stage they will have to be less based on simulation and more on expert knowledge representations. Artificial Intelligence (AI) based approaches have attempted this in the past but have not made a lasting impact, which is not surprising if one realizes that the need for early expert intervention is greatest if the problem is complex and novel. But unfortunately this is exactly where knowledge-based tools have been traditionally weak. The need for advanced tools and expert knowledge is particularly felt in those cases where it counts the most: experimental architecture that explores new frontiers in building technology. Such novel projects often pose extreme challenges through its use of very large innovative space and enclosure concepts, novel hybrid (natural and mechanical) heating and cooling concepts, a great variety of functions and occupants, different climate control regimes and more. A project of this complexity cannot be realized without involving experts at the very early stage of the design process. No architectural firm would risk relying on in-house use of designer-friendly analysis tools, because it would take a high degree of expertise to judiciously apply simplified analyses to non-routine cases (if at all sensible). It also touches on the issue of accountability for the choice of a particular analysis tool in a design problem. Modern building project partnering strategies try to deal with accountability as an integral part of team building and management. Accountability of performance analyses should be treated from the same team perspective. Designer-friendly analysis tools have typically ignored this issue by assuming that the non-expert designer will take responsibility for use of the tool. The problem with this is that to the designer, the tool is essentially a "black box", which does not make any of its applicability limitations explicit. The above assumption regarding designer responsibility seems therefore not justifiable.

Figure 1.4 reflects how designer-friendly tools are typically generated by "reduction" or "boiling down" of expert domain knowledge and expert tool functionality. The premise is that this leads to the type of tool that can be easily adopted by non-expert users in the inner design team. As there is no methodology that guides this process, tool developers use mostly heuristics in this reduction process. Ideally this mostly heuristics should be captured in explicit rules and made available to the design user as part of his conscious decisions in a design analysis scenario. It will be discussed in a later section.

The once popular research area depicted in Figure 1.4 seems to have been replaced by the opposite strategy, which is to delegate ("outsource") design analysis to domain experts and their (increasingly) complex expert tools. The latter effort concentrates on an efficient communication layer that supports the delegation of tasks and interpretation of results. Figure 1.5 shows four distinct versions of this approach that are discussed here. Whereas designer-friendly tools emphasize the import of "packaged" domain expertise into the design team, *design-integrated* tools emphasize the export of

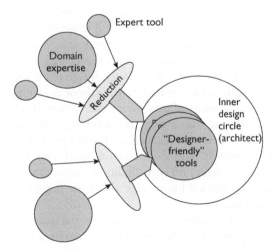

Figure 1.4 Reduction of domain knowledge in the migration of expert tools to designer-friendly tools.

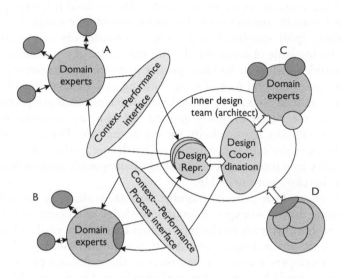

Figure 1.5 Variants of delegation of expert analysis to domain experts and their tools.

formalized analysis requests along with an explicit design context. Equally important is the import of analysis results in a form that supports better-informed rational decision-making. The basic distinction between designer-friendly tools (Figure 1.4) and design-integrated tools (Figure 1.5) is the reduction and encapsulation of domain knowledge in the first case versus enrichment and externalization of design context in the second. This has repercussions for the way that the design team operates. Instead of a tool user, the inner design team needs to become a central design manager, maintain a central design repository and act as a coordinating agent for domain experts. Variants A, B, C, and D of Figure 1.5 show different versions of how this advanced form of design evolution strategies, and especially the integration of analysis in design evolution, may be realized. The four variants differ in integration concepts and integration architecture.

Variant A represents the "classical" integration case attempted in projects like COMBINE (Augenbroe 1995). In this variant, the design context information and analysis results are exchanged between the inner design team and remote experts and their tools. The interface is data oriented, with little or no support for process management such as the management of task flow logic. When the exchange is embedded in an interoperability layer to allow easy data exchange and (automated) mappings between different views or perspectives, variant B will result. This variant uses a coordination module that controls data exchange and performs workflow management across the design team members and consultants. Contrary to variant A, the interface in variant B has access to explicit knowledge about the design analysis scenarios that are delegated to a consultant.

Variants C and D take a different approach to the team integration challenge. Variant C represents a practical partnering approach, whereas variant D is driven by deep software integration rather than on interoperability of legacy applications. In variant C a team of simulation experts is invited into the inner design team, and a high bandwidth discussion with designers is maintained throughout all stages. In McElroy and Clarke (1999) it is shown that this indeed provides the guarantees for expert simulation to be used effectively. An important condition for this variant to succeed is that it needs upfront commitment from the design team. In Variant D the emphasis is on functional and behavioral interoperability across different performance characteristics. Mahdavi *et al.* (1999) describe how this could be implemented rigorously on object level in the next generation of integrated design environments. Pelletret and Keilholz (1999) describe an interface to a modular simulation back-end with similar objectives. Both approaches have in common that they rest on the assumption that these environments will ultimately be sufficiently transparent to be accessible by members of the design team without the need for a significant reduction of domain expertise or limitations in analysis functionality. This assumption takes us back to the origin of designer-friendly tools of Figure 1.4. The four variants are expected to mature further and could possibly merge over the next 10 years.

A spin-off benefit from employing expert simulation, not always fully recognized, is the improved discipline it places on *decision-making*. As the simulation process itself is systematic, it enforces a certain level of rigor and rationality in the design team decision process. As we are progressively moving towards dispersed teams of architectural designers and analysis experts, full integration of all disciplinary tools in a collaborative design framework is the ultimate goal. This form of *"new integration"* distinguishes itself by fostering a remote engineering culture enabled by group messaging, distributed workflow management, distributed computing, supply-side component modeling and delegated simulation. Engineering design in general faces additional external challenges in the area of sustainability, resolution of conflicts across design team members, and above all performing better risk and uncertainty analyses of their performance predictions through all life cycles of the building, as described in (de Wit 2001; de Wit and Augenbroe 2001; Macdonald 2002).

The availability of function complete simulation tools is by itself no guarantee that they will play an important role in design evolution. For that, one needs to be able to guarantee that an expert simulation is called upon at the right time, and for the right design decision. Guaranteeing this is a matter of QA and adequate design process coordination. Proper coordination requires a dynamic view of all design activities, verification of their interrelatedness and anticipation of expected downstream impacts of alternative decisions. Such a dynamic view can be made explicit in a process model that (among others) captures the role of the building performance expert analysis in the design decision process. In the Architecture, Engineering and Construction (AEC) industry, the road towards a generic building representation is a long and windy one (Eastman 1999). A generic process representation does not exist either, for many obvious reasons. For one, every building project creates its own "one-off" volatile partnership and only on the highest abstraction level some established process patterns may exist that are applicable to every project. On finer granularity levels, it is commonly left to each project partnership to define and enforce the most adequate form of coordination among designers and domain experts. An important part of QA is to enforce the right type of coordination at critical decision moments in design evolution. Few unbiased empirical studies exist about the impact of building performance tools on particular design choices. Some surveys have shown that early decisions about environmental technologies are taken without adequate evidence (e.g. without being backed up by simulation) of their performance in a particular case. A recent "postmortem" analyses on a set of design projects revealed an ominous absence of the building performance analysis expert in the early stages of the design process (de Wilde *et al.* 2001). The study shows that, once the decision for a certain energy saving technology is made on the grounds of overall design considerations or particular owner requirements and cost considerations, the consultant's expertise is invoked later for dimensioning and fine-tuning. By that time the consultant is restricted to a narrow "design option space" which limits the impact of the performance analysis and follow-up recommendations. In the light of these observations, it appears a gross overstatement to attribute the majority of energy efficiency improvements in recent additions to our building stock directly to the existence of simulation tools.

In order to become an involved team player, the simulation profession needs to recognize that two parallel research tracks need to be pursued with equal vigor: (1) development of tools that respond better to design requests, and (2) development of tools that are embedded in teamware for managing and enforcing the role of analysis tools in a design project. One way to achieve the latter is to make the role of analysis explicit in a so-called project models. A project model is intended to capture all process information for a particular building project, that is, all data, task and decision flows. It contains information about how the project is managed and makes explicit how a domain consultant interacts with other members of the design team. It captures what, when and how specific design analysis requests are handed to a consultant, and it keeps track of what downstream decisions may be impacted by the invoked expertise. Design iterations are modeled explicitly, together with the information that is exchanged in each step of the iteration cycle. Process views of a project can be developed for different purposes, each requiring a specific format, depth, and granularity. If the purpose of the model is better integration, an important distinction can be made between data and process integration. Data-driven tool interoperability has been the dominant thrust of the majority of "integrated design system" research launched in the early 1990s. It is expected that process-driven interoperable systems will become the main thrust of the next decade.

1.4   New manifestations of simulation

Building simulation is constantly evolving. This section deals with three important new manifestations. As so many other engineering disciplines, the building simulation profession is discovering the WWW as a prime enabler of remote "simulation services." This will be inspected more closely in Section 1.4.1. Since the start of the development of large simulation tools, there has been the recurring desire to codify templates of standardized simulation cases. These efforts have had little success as a framework for the definition of modular simulation functionality was lacking. This may have changed with the advent of performance-based building and its repercussions for normative performance requirements. Section 1.4.2 will describe these developments and contemplate the potential consequences for the automation and codification of simulation functions. Building automation systems, sensory systems, and smart building systems (So 1999) will define the future of the "wired" building. Embedded real time control and decision-making will open a new role for embedded real time simulation. This is the subject of Section 1.4.3.

1.4.1   Web-hosted (distributed) simulation services

The web facilitates new forms of remote simulation as web-hosted services. Grand scale use of web-hosted simulation services in the building industry is not expected in the near future, but much will depend on how "Internet ready" the current simulation applications will be by the time that Application Service Providers (ASP) discover the growing market potential of e-simulation. Especially the collaboration technology providers may soon start expanding their web-hosted collaboration spaces with embedded simulation services. To understand the role of the web in various manifestations of distributed simulations such as "delegated" and collaborative simulations, it is useful to classify the various forms of concurrency in distributed simulation into four types (Fox 1996): data parallelism, functional parallelism, object parallelism, and "component parallelism". The latter form is especially suitable for distributed simulation if building components can be defined that encapsulate behavioral physics (i.e. thermal, acoustic, etc.) of building subsystems. It would stimulate shared code development and distributed simulation if this is supported by a high-level interoperability architecture allowing independent development of new components, and a component classification that is suited for distributed simulation. High-level simulation architectures have been introduced for tactical military operations where loosely coupled components such as autonomous agents in a battlefield interact at discrete events. A building is an inherently tightly coupled system, which from a distributed simulation viewpoint leads to high bandwidth data parallelism. This would not fit the distributed component supplier model. There is a need to classify the generic aspects of building systems behavior in order to define a common engineering representation consisting of building components, subcomponents and subassemblies. A major step in this direction was provided in the Neutral Model Format specification (Sahlin 1996b). Studies toward a new object class morphology could provide the necessary structure to accommodate a common engineering model, and define the essential interfaces for component coupling. In addition to the encapsulation and polymorphism concepts, messaging protocols (such as CORBA and COM) between distributed objects have been developed to optimize code-reuse and hide object implementation issues. Several studies proved that these mechanisms work well. Malkawi and Wambuagh (1999) showed how an invisible object-to-object interface allows developers to get access to a wide variety of simulation functions offered by encapsulated simulation objects.

A more immediate and low-level form of e-simulation is web-hosted simulation through an ASP business model. This variant offers web access to a "stand alone" simulation running on a remote machine. Access to the simulation application is typically enabled through a standard web browser. Although this form existed prior to the WWW revolution in traditional client–server modes, the web and Java have now created a more "natural" environment for this form of application hosting. The ASP model is rapidly expanding into engineering domains and it is only a matter of time before dedicated e-simulation services will be offered to the building industry. Web-hosted simulation has several benefits over the traditional desktop simulation. First of all, installation and maintenance on the desktops of the client organization can be avoided, whereas the latest update is always available to all users. Moreover, the developers do not need to support multiple operating platforms. The developer keeps complete control over the program and may choose to authorize its use on a case by case basis if so desired. This control may also extend to "pay per use" revenue models for commercial offerings. The opportunities to store and reuse audit-trails of user runs on the server are other benefits.

The model on the server can be made "invisible," which is useful when the user does not interact with the simulation directly but only indirectly, for instance, through a web-enabled decision support environment. An example of such application is reported by Park *et al.* (2003) in an Intranet enabled control system for smart façade technologies. The approach uses an "Internet ready" building system that plugs into the Internet making it instantly accessible from any location. In this instance, the model performs autonomous simulations in response to proposed user control interventions.

In the future, simulation may be part of an e-business service, such as the web-hosted electronic catalogue of a manufacturer of building components. Each product in the catalogue could be accompanied by a simulation component that allows users to inspect the product's response to user-specified conditions. Web hosting makes a lot of sense in this case, as manufacturers are reluctant to release the internal physical model with empirical parameters of a new product to the public. Taking it one step further, the simulation component may be part of the selling proposition and will be available to a buyer for downloading. This will for instance allow the component to be integrated into a whole building simulation. Alternatively, the component could remain on the server and be made to participate in a distributed simulation. Obviously, there are important model fidelity issues like applicability range and validation that need to be resolved before this simulation service could gain broad acceptance. Jain and Augenbroe (2003) report a slight variation on this theme. In their case, the simulation is provided as a service to rank products found in e-catalogues according to a set of user-defined performance criteria.

1.4.2   *Performance requirement driven simulation*

In the previous subsection the focus was on delivering federated or web-hosted simulation functions. Another potentially important manifestation of simulation is the automated call of simulation to assess normative building performance according to predefined metrics. This approach could become an integral part of performance-based building methods that are getting a lot of attention in international research networks (CIB-PeBBu 2003). Performance-based building is based on a set of standardized performance indicators that constitute precisely defined measures to express building performance analysis requests and their results. These performance requirements will become part of the formal statement of requirements (SOR) of the client. They are expressed in quantified performance indicators (PIs). During design evolution, the domain expert analyzes and assesses the design variants against a set of predefined PIs addressed in the SOR, or formulated during the design analysis dialogue, that is, expressed by the design team as further refinements of the clients requirements. In this section a theoretical approach is discussed which could effectively underpin performance-based design strategies by performance metrics that are based on simulation. At this point of time no systems exist to realize this, although an early try is reported by Augenbroe *et al.* (2004).

The role of simulation during this process could be centered around a (large) set of predefined PIs. Every PI is an unambiguously defined measure for the performance that a particular building system has towards a given function of that system or a larger system of which it is a part. The quantification of a PI requires a (virtual) experiment on the system under study. Every PI comes with exactly one experiment, which describes exactly one system type, a particular experiment on that system, and a way to aggregate the results of the experiment. One should keep in mind that a particular function of a building system may have multiple ways of measuring the performance of that system towards that function. Each different way of defining an aspect or measuring method of that aspect leads to one unique PI.

The quantification of a PI is linked to a precisely defined experiment, which can be conducted in whatever form as long as it generates the output states of the object that need to be observed, analyzed and aggregated for the PI quantification process. In fact, any form of aggregation that leads to a relevant measure of the behavior of the building system is a candidate for a PI. The main qualification test on a PI is that it can be unambiguously linked to a desired function of the object, can be reproduced in repeated experiments and is meaningful to a design decision.

This provides an "artificial" framework for "interpretation-free" simulations with the biggest proclaimed benefit that such simulations could be easily automated as simulation agents in performance dialogues. This would not only underpin dialogues during design evolution but also in commissioning and in-use building performance monitoring (e.g. for continuous commissioning). Although many are trying to codify performance indicators into standards, the link to a "pre-configured" set of simulation functions seems as yet unlikely. After all, in order to deliver a reproducible PI value, its associated analysis function (the "experiment") needs to be executed in an objective way hence its outcome must be independent of the expert and his tools. This demand is necessary as it creates independence of the performance dialogue from any particular simulation software. If this demand cannot be met, the result of the design analysis dialogue becomes dependent on the particular software user combination. Although this is clearly unwanted, it is not totally unavoidable. After all, even under exactly defined circumstances, different tools provide slightly different answers due to differences in modeling assumptions, solver accuracy, etc. To support a PI-based dialogue adequately by simulation tools, one would have to "accredit" a software tool for a given set of PIs. Different types of experiments will require different "accreditation" methods. In a real experiment, it is not always possible to control the input variables, whereas in a virtual experiment, one has only limited control over the process assumptions and schematizations that the software deploys for its native representation of the experiment. A thought experiment seems even more subject to "biases" as too many judgment calls raise doubts on the reproducibility across experts. It should be kept in mind, however, that the use of virtual experiments is the very basis of the current design analysis practices. Indeed, in order to support the expression of functional requirements by the design team some behavioral experiment must be assumed and known to both the design team and the experts. The introduction of the PI concept does not radically change the fundamentals of this design analysis dialogue, it just makes it more precise through formal specification and quantification of its elements. Use of a standard PI guarantees that the purpose of the experiment (the performance) is understood by all those engaged in the dialogue. This in itself is a tremendous step forward from the current situation, where too much depends on interpretation of an unstructured dialogue, creating a "service relationship" with the consultant which can be inefficient.

Adopting the above would force a new and close look at the current simulation tools, none of which have been developed with a set of precise PI evaluations in mind. Rather they were developed as a "free style" behavioral study tool, thus in fact creating a semantic mismatch between its input parameters and the use for specific performance assessments. It remains an open question whether the performance-based building community will put enough pressure on simulation tool providers to define a set of standard PI quantifications as explicit analysis functions offered by the tools.

1.4.3   Real time building simulation

There is an ongoing trend to embed real time simulation in automation, control and warning system. This trend is fed by an increasing need to inform real time decision-making by sensory (actual) and what-if (predicted) state information. It is therefore necessary to develop new interaction modes for human decision-makers and automated systems that respond to varying degrees of time criticalness and varying needs of granularity. This leads to a heterogeneous set of adaptable and scalable simulation tools that are capable of responding to the identified needs of stakeholders. This will also include tightly coupled building and control system simulations enabling the design of adaptive and predictive control strategies. For instance, in the case of emergency response to chemical and biological hazards, tools should be able to perform uncertainty analyses for operational decision-making based on assumed model and parameter uncertainties and sensor error. The interaction modes that support these needs will adapt to new multimedia devices and display technologies. Current interaction paradigms are quickly going to be replaced by new developments such as the Power Browser (Buyukkokten *et al.* 2000), iRoom (Liston *et al.* 2001), and Flow Menu (Guimbretière and Winograd 2000). Wearable computing (Starner 2002) and augmented reality (Malkawi and Choudhury 1999; MacIntyre 2000) provide an appropriate starting point towards radically new ways to interact with embedded simulation.

Control systems are becoming intelligent and model based, that is, control logic is only partly predetermined and embedded in hardware. The major part of the control logic is being defined declaratively embedded in behavioral models of the object under control (Loveday *et al.* 1997). The controlled behavior can be inspected and anticipated before a particular control action is executed. The underlying model has to respond to the needs to (1) analyze and judge plausibility of incoming data by making inferences, (2) adapt model parameters through continuous calibration, and (3) find optimal control and response measures in any given situation based on determination of best action. Simulation models now play a major role in the design of control strategies and in technical system design. New theories in model-based control take the notion of "minimalistic" simulation models a step further by constructing them only from minimal physical knowledge and calibrating them on the set of anticipated control actions (Decque *et al.* 2000; Tan and Li 2002). Simulation tools will increasingly be embedded in control systems design in order to identify the critical system parameters for control systems to assist human decision-making. This requires tight coupling of building automation and control systems with embedded adaptive and predictive simulation-based control strategies capable to reason about uncertain data and respond to occupant interventions and random events. Also, the interpretation of sensor data will be simulation-driven instead of static rules driven. All of the above will become an integral component of the next generation of simulation-driven commercial building control systems.

1.5   Final remarks

In less than thirty years building simulation tools have evolved from primitive equation solvers to validated large-scale software codes with a large user base. It is to be expected that extensions of building simulation tools will be driven by the need to issue better quality control over the performance assessment dialogue with other members of the project. Although there remain weaknesses and gaps in tool functionality, the more immediate challenge is to better integrate simulation in all phases of the building process.

Object-oriented frameworks respond to the needs for shared codevelopment by leveraging proven software design. The approach should be based on a reusable component-based architecture that can be extended and customized to meet future application requirements. Such a high-level architecture is still elusive in the building industry although important building blocks are already available. The Internet is the natural environment for distributed simulation but the building performance research community faces the uphill task of developing a common engineering representation that is capable of providing the high-level architecture for component sharing. With this architecture in place, the Web could act as a catalyst for top down development of interoperable components, and simulation could become an integral component in teamware for collaboration in design and engineering teams. With this in mind, building performance experts should proactively engage in the deployment of a new breed of team ware for project management, as this will ultimately enable the profession to control its own destiny in a project team setting through proper input and role specification of the role of the building performance expert at project definition.

With the advent of these resources, it may ultimately be appropriate to enforce a formal agreement between design team and building simulation expert, concerning the model assumptions that underlie a delivered design analysis. Model specifications that are suitable for such formal agreement do not exist in current practice. Research in this area should deal with certification and expert calibration based on approaches that use uncertainty and risk analysis.

This range of physical models continues to expand in dedicated transport model development, for example, in the fields of mold growth (Holm *et al.* 2003), fire/smoke dynamics, occupant comfort models, and ongoing attempts to create reliable models in the complex field of small particle (especially bioaerosol) movements in indoor spaces (Liu and Nazaroff 2002; Loomans *et al.* 2002; Sextro *et al.* 2002).

New manifestations of simulations are likely to appear in web-hosted service provision, performance-based frameworks and building automation and control systems. All three manifestations are nascent and hold great promise for a future in which running a simulation only requires an Internet browser, provides choice among a set of "exposed" modular standardized functions and is responsive to real time decision-making.

References

Augenbroe, G.L.M. (1986). "Research-oriented tools for temperature calculations in buildings." In *Proceedings of 2nd International Conference on System Simulation in Buildings*, Liege.

Augenbroe, G. (1995). "COMBINE 2 Final Report." CEC Publication, Brussels.

Augenbroe, G., Malkawi, A., and de Wilde, P. (2004). "A workbench for structured design analysis dialogues." *Journal of Architectural and Planning Research* (Winter).

Bazjanac, V. and Crawley, D. (1999). "Industry foundation classes and interoperable commercial software in support of design of energy-efficient buildings." In *Proceedings of Building Simulation '99*, Sixth International IBPSA Conference, Kyoto, September, Paper B-18.

Björsell, N., Bring, A., Eriksson, L., Grozman, P., Lindgren, M., Sahlin, P., Shapovalov, A., and Vuolle, M. (1999). "IDA indoor climate and energy." In *Proceedings of Building Simulation '99*, Sixth International IBPSA Conference, Kyoto, September, Paper PB-10.

Buyukkokten, O., Garcia-Molina, H., Paepcke, A., and Winograd, T. (2000). "Power browser: efficient web browsing for PDAs." Human–Computer Interaction Conference 2000 (CHI 2000). The Hague, The Netherlands, April 1–6.

CIB-PeBBu (2003). *Thematic Network in Performance Based Building*. Rotterdam, Netherlands, http://www.pebbu.nl/

Clarke, J.A. (1999). "Prospects for truly integrated building performance simulation." In *Proceedings of Building Simulation '99*, Sixth International IBPSA Conference, Kyoto, September, Paper P-06.

Clarke, J.A. (2001). *Energy Simulation in Building Design*, Butterworth Heinemann, Oxford.

Clarke, J.A. and Hensen, J.L.M. (2000). "Integrated simulation for building design: an example state-of-the-art system." In *Proceedings International Conference Construction Information Technology 2000* (CIT2000), CIB-W78/IABSE/EG-SEA-AI, Icelandic Building Research Institute, Reykjavik, Vol. 1, pp. 465–475.

Crawley, D.B., Pedersen, C.O., Liesen, R.J., Fisher, D.E., Strand, R.K., Taylor, R.D., Lawrie, L.K., Winkelmann, F.C., Buhl, W.F., Erdem, A.E., and Huang, Y.J. (1999). "ENERGYPLUS, A new-generation building energy simulation program." In *Proceedings of Building Simulation '99*, Sixth International IBPSA Conference, Kyoto, September, Paper A-09.

Curlett, B.P. and Felder, J.L. (1995). *Object-oriented Approach for Gas Turbine Engine Simulation*. NASA TM-106970.

Deque, F., Ollivier, F., and Poblador, A. (2000). "Grey boxes used to represent buildings with a minimum number of geometric and thermal parameters." *Energy and Buildings*, Vol. 31, p. 29.

De Wilde P., van der Voorden, M., Brouwer, J., Augenbroe, G., and Kaan, H. (2001). "The need for computational support in energy-efficient design projects in The Netherlands." In *Proceedings of BS01*, 7th International IBPSA Conference, Rio, August, pp. 513–519.

DOE (2003). "US Department of Energy. Building Energy Software Tool Directory." Washington, USA. Available from: http://www.eren.doe.gov/buildings/tools_directory/

Dupagne, A. (ed.) (1991). "ESPRIT-CIB Exploratory action 'Computer Integrated Building'," CEC-report DG XIII.

Eastman, C.M. (1999). *Building Product Models: Computer Environments Supporting Design and Construction*. CRC Press, New York.

Elmqvist, H., Mattson, S., and Otter, M. (1999). "Modelica—a language for physical system modeling, visualization and interaction." In *IEEE Symposium on Computer-Aided Control System Design*, CACSD'99, Hawaii.

Fishwick P.A. and Zeigler, B.P. (1992). "A multimodel methodology for qualitative model engineering." *ACM Transactions on Modeling and Computer Simulation*, Vol. 12, pp. 52–81.

Fox, G.C. (1996). "An application perspective on high-performance computing and communications." Technical Report SCCS-757, Syracuse University, NPAC, Syracuse, NY, April.

Fujii, H. and Tanimoto, J. (2003). "Coupling building simulation with agent simulation for exploration to environmental symbiotic architectures." In Eighth IBPSA Conference on Building Simulation, Eindhoven. Godfried Augenbroe and Jan Hensen (eds), August, pp. 363–370.

Guimbretière, F. and Winograd, T. (2000). "FlowMenu: combining command, text, and data entry." UIST'00, ACM, pp. 213–216.

Holm, A., Kuenzel, H.M., and Sedlbauer, K. (2003). "The hygrothermal behavior of rooms: combining thermal building simulation and hygrothermal envelope calculation." In Eighth IBPSA Conference on Building Simulation, Eindhoven. Godfried Augenbroe and Jan Hensen (eds), August, pp. 499–505.

IBPSA (2003). *Proceedings of First-eight Building Simulation Conferences*, on CD, IBPSA (1989–2003).

Jain, S. and Augenbroe, G. (2003). "A Methodology for Supporting Product Selection from e-catalogues." In Robert Amor and Ricardo Jardim-Gonçalves (eds), Special Issue on eWork and eBusiness. ITCON, November.

Liston, K., Fischer, M., and Winograd, T. (2001). "Focused sharing of information for multi-disciplinary decision making by project teams." ITCon (*Electronic Journal of Information Technology in Construction*), Vol. 6, pp. 69–81.

Liu, D.L. and Nazaroff, W.W. (2002). "Particle penetration through windows." In *Proceedings Indoor Air 2002*, pp. 862–867.

Loomans, M.G.L.C., Bluyssen, P.M., and Ringlever-Klaassen, C.C.M. (2000). "Bioaerosols-where should one measure them in a room?" In *Proceedings Indoor Air 2002*, pp. 443–448.

Loveday, D.L., Kirk, G.S., Cheung, J.Y.M., and Azzi, D. (1997). "Intelligence in buildings: the potential of advanced modeling." *Automation in Construction*, Vol. 6, pp. 447–461.

Macdonald, I.A. (2002). "Quantifying the Effects of Uncertainty in Building Simulation." PhD thesis. University of Strathclyde, July.

McElroy, L. and Clarke, J.A. (1999). "Embedding simulation within the energy sector business." In *Proceedings of the Building Simulation '99*, Sixth International IBPSA Conference, Kyoto, pp. 262–268.

MacIntyre, B. (2000). "Context-aware personal augmented reality." Position paper at CHI'00 Workshop on Research Directions in Situated Computing, The Hague, The Netherlands.

Mahdavi, A. (2001). "Distributed multi-disciplinary building performance computing." In *Proceedings of the 8th Europia International Conference Delft*, R. Beheshti (ed.), The Netherlands, pp. 159–170.

Mahdavi, A., Mustafa, A.I., Matthew, P., Ries, R., Suter, G., and Brahme, R. (1999). "The architecture of S2." In *Proceedings of the Building Simulation 99*, Sixth International IBPSA Conference, Kyoto, September, Paper A-38.

Malkawi, A. and Choudhary, R. (1999). "Visualizing the sensed environment in the real world." *Journal of the Human-Environment Systems*, Vol. 3, No. 1, pp. 61–69.

Malkawi, A. and Wambaugh, J. (1999). "Platform independent simulations: thermal simulation as an object". In *Proceedings of the 6th International Building Simulation Conference*, Kyoto, Japan.

Park, C., Augenbroe, G., Sadegh, N., Thitisawat, M., and Messadi, T. (2003). "Occupant response control systems of smart façades." In Eighth IBPSA Conference on Building Simulation, Eindhoven. Godfried Augenbroe and Jan Hensen (eds), August, pp. 1009–1016.

Pelletret, R. and Keilholz, W. (1999). "Coupling CAD tools and building simulation evaluators." In *Proceedings of the Building Simulation '99*, Sixth International IBPSA Conference, Kyoto, Japan, 13–15, pp. 1197–1202.

Sahlin, P. (1996a). "Modelling and simulation methods for modular continuous systems in buildings." Doctoral Dissertation KTH, Stockholm, Sweden (also available at http://www.brisdata.se/ida/literature.htm).

Sahlin, P. (1996b). *NMF Handbook, An Introduction to the Neutral Model Format*, NMF Version 3.02, ASHRAE RP-839. Report from Building Sciences, KTH, Stockholm.

Schmidt, D.C. (1997). "Applying design patterns and frameworks to develop object-oriented communications software." In P. Salus (ed.), *Handbook of Programming Languages*. Vol. I, MacMillian Computer Publishing.

Sextro, R.G., Lorenzetti, D.M., Sohn, M.D., and Thatcher, T.L. (2002). "Modeling the Spread of Anthrax in Buildings." In *Proceedings Indoor Air 2002*, pp. 506–511.

So, A.T. (1999). *Intelligent Building Systems*. Kluwer Academic, Boston.

Sowell, E.F. and Haves, P. (1999). "Numerical performance of the SPARK graph-theoretic simulation program." In *Proceedings of Building Simulation 99*, Sixth International IBPSA Conference, Kyoto, September, Paper A-05.

Starner, T. (2002). "Thick clients for personal wireless devices." *IEEE Computer*, Vol. 35, No. 1, pp. 133–135.

Tan, K.C. and Li, Y. (2002). "Grey-box model identification via evolutionary computing." *Control Engineering Practice*, Vol. 10, pp. 673–684.

Tang, D. and Clarke, J.A. (1993). "Application of the object oriented programming paradigm to building plant system modelling." In *Proceedings of Building Simulation '93*, Third International IBPSA Conference, Adelaide.

de Wit, M.S. (2001). "Uncertainty in predictions of thermal comfort in buildings." Doctoral Dissertation, TU Delft, June.

de Wit, S. and Augenbroe, G. (2001). "Uncertainty analysis of building design evaluations." In *Proceedings of BS01*, 7th International IBPSA Conference, Rio, August, pp. 319–326.

Chapter 2

Uncertainty in building simulation

Sten de Wit

2.1   Introduction

Building simulation facilitates the assessment of the response of a building or building component to specified external conditions by means of a (computer) model. It is an instrument, which is exceptionally suitable to answer "what if"-type questions. "What would happen if we would make this design alteration?" "What would be the effect of this type of retrofit?" "How would the building respond to these extreme conditions?" This type of questions typically arise in a decision-making context, where the consequences of various alternative courses of action are to be assessed.

Commonly these consequences can only be estimated with some degree of uncertainty. This uncertainty may arise from a variety of sources. The first source is a lack of knowledge about the properties of the building or building component. This lack of knowledge is most evident when the simulations concern a building under design. But even when the object of study is an existing building and its properties can be measured in theory, practical limitations on time and money will generally be prohibitive for a precise specification of the building properties.

Moreover, in addition to the lack of knowledge about the building itself, several external factors, which drive the building's response of interest, may not be precisely known. Finally, the complexity of the building commonly makes it necessary to introduce simplifications in the computer simulation models. Together with the lack of information about the building and the external factors it will be exposed to, these simplifications lead to uncertainty in the simulation outcome.

In practical applications of building simulation, explicit appraisal of uncertainty is the exception rather than the rule and most decisions are based on single-valued estimates. From a conceptual point of view, this lack of concern for uncertainty is surprising. If we consider building simulation as an instrument, which aims to contribute to decision-makers' understanding and overview of the decision-problem, it seems natural that uncertainties are assessed and communicated.

From a practical perspective, though, the lack of focus on uncertainty is quite natural. In current practice, building simulation is commonly performed with commercially available tools. Such tools facilitate the modeling and simulation of complex building systems within the limitations on time and money that apply in practical situations. However, the tools provide virtually no handles to explore and quantify uncertainty in the assessments.

First, no information is supplied about the magnitudes of the various uncertainties that come into play. Libraries with data on, for example, material properties and model parameters, which are included in almost all simulation tools, specify default or "best" values, but lack information on the spread in these values. Second, with the exception of one or two, none of these tools offer methods to carry out a systematic sensitivity analysis or to propagate uncertainty. Finally, the possibilities to selectively refine or simplify model aspects are limited in most simulation environments.

In the building simulation research field, several studies have been dedicated to uncertainty in the output of building simulations and the building performance derived from these outputs. Report of the most relevant research can be found in Lomas and Bowman (1988), Clarke *et al.* (1990), Pinney *et al.* (1991), Lomas and Eppel (1992), Lomas (1993), Martin (1993), Fürbringer (1994), Jensen (1994), Wijsman (1994), Rahni *et al.* (1997), de Wit (1997b, 2001), MacDonald *et al.* (1999), MacDonald (2002, 2003), de Wit and Augenbroe (2002). These studies indicate that adequate data on the various uncertainties that may contribute to the uncertainty in building performance is limited. Among these, uncertainties related to natural variability, which can sensibly be quantified on the basis of statistical analysis such as spread in, for example, material properties and building dimensions are relatively well covered. Modeling uncertainties, though, and other uncertainties that cannot be comprehensively derived from observed relative frequencies, have received only limited attention, and usually only on an ad hoc basis. Although several of the studies have focused on a comparison of techniques for sensitivity analysis and propagation of uncertainty, these techniques have hardly pervaded the mainstream tools for building simulation. Virtually no concern is given to the question how quantitative uncertainty can be used to better-inform a design decision.

This chapter illustrates how uncertainties in building simulations can be addressed in a rational way, from a first exploration up to the incorporation of explicit uncertainty information in decision-making. Most attention is given to those issues, which have been sparsely covered in the building simulation literature, that is modeling uncertainties and decision-making under uncertainty. To keep the discussion of these issues as tangible as possible, this chapter is constructed around a specific case.

Section 2.2 presents an outline of the case. Subsequently, in Section 2.3 the main issues of uncertainty analysis are discussed and applied to the case. Section 2.4 shows how the uncertainty analysis can be refined, guided by the findings of the analysis in Section 2.3. A demonstration of how the compiled information on uncertainties can be constructively used in a decision analysis is elaborated in Section 2.5. Finally, Section 2.6 concludes with summary and outlook.

2.2   Outline of the case

Uncertainties in building simulations are especially relevant when decisions are made on the basis of the results. Hence, a decision-making problem is selected as a suitable case. The context is a (advanced) design stage of an office building in The Netherlands. In the moderate climate it is possible to make naturally ventilated buildings, which are comfortable in summer. Hence, the choice to either or not install a cooling plant is a common design issue. This decision-problem will be addressed here. In the next section, the main characteristics of the office building and its immediate environment are outlined. Subsequently, the decision-problem is described in Section 2.2.2.

2.2.1 Building and its environment

The context of the decision-making problem is an advanced design stage of a four-story office building in a suburban/urban environment in The Netherlands. Figure 2.1 shows a front view of the office building with its main dimensions.

In summer, cantilever windows in the long façades can be opened to control the indoor temperatures. The building is designed in such a way that the main ventilation mechanism is cross-ventilation, driven by local wind pressure differences between the opposite long façades. These wind pressures are sensitive to the topology of the environment of the building. An outline of the environment is given in Figure 2.2. The upper half of the area shows a typical urban setting. The lower half is left void, with exception of the embankment of the roadway. This is a large open space in the

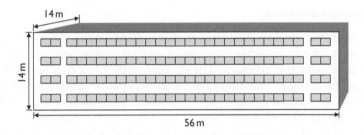

Figure 2.1 Schematic view of the office building with its main dimensions.

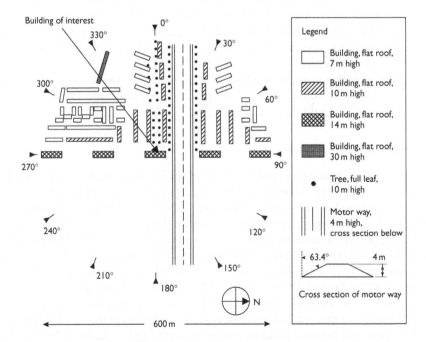

Figure 2.2 Schematic layout of the building and its environment.

otherwise urban environment. For later reference (Sections 2.3.3 and 2.4.2), azimuth angles relative to the west are plotted.

Without cooling plant in the building it will be most difficult to maintain acceptable climatic conditions in the spaces on the top floor, especially those oriented to the east. Hence, as a first step in the assessment of the performance of the building with respect to indoor climate, the thermal conditions in one of these office spaces will be studied by means of building simulation.

Actual development and execution of the building simulation model involves much more information about the design specifications and scenario. However, as this information is not relevant to the argument in this chapter, it will be omitted. Those who are interested in the details are referred to de Wit (2001), where this case is fully documented.

2.2.2   Decision-problem

In this example, we consider the situation that only two actions are of concern to the decision-maker, that is, he either integrates a modest cooling system in the design, or he doesn't and saves the expenses. The decision-maker has two (conflicting) objectives to guide his actions: "maximize the future occupants' satisfaction with the (thermal aspects of the) indoor climate" and "minimize investment cost". To measure the achievement on the first objective he uses the TO, a performance indicator for thermal comfort. The TO, commonly used in The Netherlands, expresses the number of office hours per year that the operative indoor temperature exceeds 25.5 °C under a specific scenario, that is, external climate conditions and occupation profile. We assume that the investment cost associated with both actions is known without substantial uncertainty. The TO-indicator will be equal to 0 in case the cooling is installed, as this system will be dimensioned to achieve this. The possibility of failure of this system is not considered here. The investment cost of the system is set to 400×10^3 monetary units.

The question in this example is which of the two actions the decision-maker should choose. This depends on the value, of the performance indicator TO under action 1. To assess this value, building simulation is deployed.

2.2.3   Simulation approach

A variety of methods and tools exist to perform the building simulation. In this chapter a "consensus approach" based on approaches from Clarke (1985) and Augenbroe (1986), and reflected in many mainstream tools (DOE 2003) has been chosen. A recent trend analysis of building simulation techniques can be found in Augenbroe (2000). The consensus approach is based on a nodal space discretization (i.e. based on the finite element or finite difference technique). The spatial nodes contain the discrete values of the state variables. Ventilation flow rates are modeled by a basic network approach (Liddament 1986; Feustel 1990). The modeling technique leads to a space discretized system of ode's and algebraic equations in the state variables, which are subsequently solved by numerical time integration. The actual simulations in the case study have been executed with two thermal building modeling tools, ESP-r (ESRU 1995) and BFEP (Augenbroe 1982–1988).

This completes the outline of the case that will be used as an example throughout this chapter. The case concerns a decision-problem, the analysis of which requires assessment of the consequences of a particular action in terms of a building performance indicator. This assessment involves uncertainty. The analysis of this uncertainty is the topic of the next section.

2.3 Uncertainty analysis

2.3.1 Introduction

Uncertainty may enter the assessment of building performance from various sources. First, the design specifications do not completely specify all relevant properties of the building and the relevant installations. Instead of material properties, for instance, material types will commonly be specified, leaving uncertainty as to what the exact properties are. Moreover, during the construction of the building, deviations from the design specifications may occur. This uncertainty, arising from incomplete specification of the system to be modeled will be referred to as *specification* uncertainty.

Second, the physical model development itself introduces uncertainty, which we will refer to as *modeling* uncertainty. Indeed, even if a model is developed on the basis of a complete description of all relevant building properties, the introduction of assumptions and the simplified modeling of (complex) physical processes introduce uncertainty in the model.

Third, numerical errors will be introduced in the discretization and simulation of the model. We assume that this *numerical* uncertainty can be made arbitrarily small by choosing appropriate discretization and time steps. Hence, this uncertainty will not be addressed here.

Finally, uncertainty may be present in the *scenario*, which specifies the external conditions imposed on the building, including for example outdoor climate conditions and occupant behavior. The scenario basically describes the experiment, in which we aim to determine the building performance.

To quantitatively analyze uncertainty and its impact on building performance, it must be provided with a mathematical representation. In this study, uncertainty is expressed in terms of probability. This representation is adequate for the applications of concern in this work and it has been studied, challenged, and refined in all its aspects.

Moreover, in interpreting probability, we will follow the subjective school. In the subjective view, probability expresses a degree of belief of a single person and can, in principle, be measured by observing choice behavior. It is a philosophically sound interpretation, which fulfills our needs in decision analysis.

It should be mentioned, however, that in the context of rational decision-making, one subjective probability is as good as another. There is no rational mechanism for persuading individuals to adopt the same degree of belief. Only when observations become available, subjective probabilities will converge in the long run. However, the aim of uncertainty analysis is not to obtain agreement on uncertainties. Rather, its purpose is to explore the consequences of uncertainty in quantitative models. Discussions and background on the interpretation of probability can be found in for example Savage (1954), Cooke (1991), and French (1993).

Now that we have discussed the various sources of uncertainty and decided to use probability to measure uncertainty, the question is now how to analyze the uncertainty in building performance arising from these sources in terms of probability (distributions). The principles of uncertainty analysis are introduced in Section 2.3.2. In the subsequent sections, a crude uncertainty analysis is elaborated in the context of the case described in Section 2.2: uncertainties in model parameters and scenario are estimated and propagated through the simulation model. Sections 2.3.3, 2.3.4, and 2.3.5 deal with these issues respectively. If uncertainty is found to be of significance, it is directive for further steps to find out which parameters are the predominant contributors. Sensitivity analysis is a useful technique to further this goal. It is discussed in Section 2.3.6.

2.3.2   Principles of uncertainty analysis

2.3.2.1   Introduction

We start from a process scheme of building performance assessment as shown in Figure 2.3.

The figure is also a process scheme for uncertainty analysis. The difference with the deterministic case is that model parameters may now be uncertain variables. This implies that the process elements are more complex. For instance, parameter quantification now requires not (only) an assessment of a point estimate, but (also) an assessment of the uncertainty. Moreover, in the presence of uncertainty, model evaluation is a process, which propagates uncertainty in scenario and parameters through the model into the model output. Furthermore, the scope of the sensitivity analysis is extended. Besides the sensitivities, the importance of the variables can also be assessed now. The term "importance" is used here to express the relative contribution of a variable (or set of variables) to the uncertainty in the model output, that is, the building performance.

Especially in the presence of uncertainty, it is better to assess performance in a cyclic rather than a linear approach. Proper assessment of uncertainties in parameters

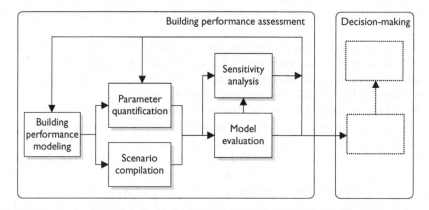

Figure 2.3 Process scheme of building performance assessment as input to decision-making.

and inputs may be a formidable task. By starting with crude estimates, and deciding on selective refinement to those variables that really matter, the problem becomes tractable. This approach is applied here: in this section a crude uncertainty analysis will be presented, which will be refined at specific aspects in Section 2.4. However, before embarking on the crude uncertainty analysis, the principles of uncertainty analysis will first be explained in more detail in the next subsections. Sections 2.3.3 and 2.3.4 address the assessment of uncertainties in model and scenario parameters. Subsequently, Section 2.3.5 deals with the propagation of uncertainty through (building simulation) models, whereas in Section 2.3.6 sensitivity analysis is introduced.

2.3.2.2 Assessment of uncertainty in parameters

In cyclic assessment, the first stage is crude, and uses existing information. For each parameter, probability distribution plus possibly statistical dependencies between parameters is evaluated. The first step is to assess plausible ranges, assign interpretation in terms of probability (e.g. 90% confidence interval), and assume common type of probability distribution. We will assume dependencies/correlations to be either absent or complete, followed by refinement in later cycles of analysis if desirable.

Synthesis of information from, for example, the literature, experiments, model calculations, rules of thumb, and experience. Nothing new, but focus is now not only on a "best" estimate, but also on uncertainty. Uncertainty may become apparent from, for example, spread in experimental results and calculation results, conflicting information, or lack of data. There is no general rule on how to quantify this uncertainty. Examples are given in Section 2.3.3.

2.3.2.3 Propagation of uncertainty

Once the uncertainty in the model parameters is quantified, the resulting uncertainty in the model output is to be assessed. This process is referred to as the propagation of the uncertainty. A variety of propagation techniques can be found in the literature, for example, in Iman and Helton (1985), Janssen *et al.* (1990), McKay (1995), MacDonald (2002), Karadeniz and Vrouwenvelder (2003). It is outside the scope of this chapter to give an overview of the available techniques. We will limit the discussion to the criteria to select a suitable method. Moreover, an appropriate technique will be described to propagate the uncertainty in the example case.

SELECTION CRITERIA FOR A PROPAGATION TECHNIQUE

The first question in the selection process of a propagation technique is: what should the propagation produce? Although propagation of uncertainty ideally results in a full specification of the (joint) probability distribution over the simulation output(s), this is neither feasible nor necessary in most practical situations. Commonly, it is sufficient to only calculate specific aspects of the probability distribution, such as mean and standard deviation or the probability that a particular value is exceeded. Dependent on the desired propagation result, different techniques may be appropriate.

A second criterion for the selection of a technique is the economy of the method in terms of the number of model evaluations required to obtain a sufficiently accurate result. In fact, economy is one of the important motives in the ongoing development of new propagation techniques. More economic or efficient methods often rely on specific assumptions about the model behavior such as linearity or smoothness in the parameters. To obtain reliable results with such methods, it is important to verify whether these assumptions hold for the model at hand.

In practical situations, an additional aspect of interest is commonly the ease and flexibility to apply the method to the (simulation) model.

SELECTED TECHNIQUE IN THE CASE EXAMPLE

The purpose of the propagation in the example case is to estimate the mean and standard deviation of the model output (building performance), and to obtain an idea of the shape of the probability distribution. The most widely applicable, and easy to implement method for this purpose is Monte Carlo simulation.[1] It has one drawback: it requires a large number of model evaluations. In the example case this is not a big issue. Obviously, if computationally intensive models are to be dealt with (e.g. Computational Fluid Dynamics, CFD), this will become an obstacle.

In the example case, however, we will use Monte Carlo (MC) simulation. To somewhat speed up the propagation, a modified Monte Carlo technique will be applied, that is, Latin Hypercube Sampling (LHS). This is a stratified sampling method. The domain of each parameter is subdivided into N disjoint intervals (strata) with equal probability mass. In each interval, a single sample is randomly drawn from the associated probability distribution. If desired, the resulting samples for the individual parameters can be combined to obtain a given dependency structure. Application of this technique provides a good coverage of the parameter space with relatively few samples compared to simple random sampling (crude Monte Carlo). It yields an unbiased and often more efficient estimator of the mean, but the estimator of the variance is biased. The bias is unknown, but commonly small. More information can be found in, for example, McKay *et al.* (1979), Iman and Conover (1980), and Iman and Helton (1985).

2.3.2.4 *Sensitivity analysis*

In the context of an uncertainty analysis, the aim of a sensitivity analysis is to determine the importance of parameters in terms of their contribution to the uncertainty in the model output. Sensitivity analysis is an essential element in a cyclic uncertainty analysis, both to gain understanding of the makeup of the uncertainties and to pinpoint the parameters that deserve primary focus in the next cycle of the analysis.

Especially in first stages of an uncertainty analysis only the ranking of parameter importance is of interest, rather than their absolute values. To that purpose, crude sensitivity analysis techniques are available, which are also referred to as parameter screening methods.

SELECTION CRITERIA FOR A PARAMETER SCREENING TECHNIQUE

Techniques for sensitivity analysis and parameter screening are well documented in the literature, for example, in Janssen *et al.* (1990), McKay (1995), Andres (1997),

Kleijnen (1997), Reedijk (2000), and Saltelli *et al.* (2000). We will not give an overview here, but restrict the discussion to the criteria for selection. Moreover, a technique for use in the example case is selected and explained.

The first issue in the selection of a sensitivity analysis technique concerns the definition of importance. Loosely stated, the importance of a parameter is its (relative) contribution to the uncertainty in model output. This is a clear concept as long as the output uncertainty can be (approximately) considered as the sum of uncertainty contributions that are attributable to individual parameters. However, if parameter interactions come into play, this concept needs refinement. This is even more the case when dependencies between variables are to be considered. As most sensitivity analysis techniques are centered around a specific interpretation of "importance", it is necessary to reflect which interpretation best fits the problem at hand.

Other criteria in the selection of a sensitivity analysis technique are very similar to the criteria for propagation techniques.

SELECTED TECHNIQUE IN THE CASE EXAMPLE

In this analysis, the factorial sampling technique as proposed by Morris (1991) has been used. In an earlier analysis (de Wit 1997c), this technique was found to be suitable for application with building models. It is economical for models with a large number of parameters, it does not depend on any assumptions about the relationship between parameters and model output (such as linearity) and the results are easily interpreted in a lucid, graphical way. Moreover, it provides a global impression of parameter importance instead of a local value. Thus, the effect of a parameter on the model output is assessed in multiple regions of the parameter space rather than in a fixed (base case) point in that space. This feature allows for exploration of non-linearity and interaction effects in the model.

A possible drawback of the method is that it does not consider dependencies between parameters. In situations where a lot of information on the uncertainty or variability of the parameters is available this might be restrictive, but in this crude analysis this is hardly the case.

In this method, the sensitivity of the model output for a given parameter is related to the *elementary effects* of that parameter. An elementary effect of a parameter is the change in the model output as a result of a change Δ in that parameter, while all other parameters are kept at a fixed value. By choosing the variation Δ for each parameter as a fixed fraction of its central 95% confidence interval, the elementary effects become a measure of parameter importance.

Clearly, if the model is nonlinear in the parameters or if parameters interact, the value of the elementary effect of a parameter may vary with the point in the parameter space where it is calculated. Hence, to obtain an impression of this variation, a number of elementary effects are calculated at randomly sampled points in the parameter space.

A large sample mean of the elementary effects for a given parameter indicates an important "overall" influence on the output. A large standard deviation indicates an input whose influence is highly dependent on the values of the parameters, that is, one involved in interactions or whose effect is nonlinear.

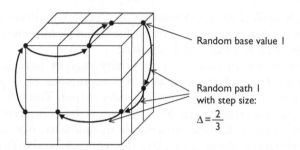

Figure 2.4 An illustration of the procedure to assess two samples of the elementary effect of each parameter. The illustration is for a three-dimensional parameter space with a 4-level grid ($k = 3, p = 4, r = 2$).

Hence, an overview of the output of the sensitivity analysis can be obtained from a graph in which sample mean and standard deviation of the elementary effects are plotted for each of the parameters.

Technically, the procedure is implemented as follows. Each of the k model parameters is scaled to have a region of interest equal to [0, 1]. The scaled k-dimensional parameter vector is denoted by $\underline{x}$. For each parameter, the region of interest is discretized in a p-level grid, where each x_i may take on values from $\{0, 1/(p-1), 2/(p-1), \ldots, 1\}$.

The elementary effect d of the ith input is then defined by

$$d_i(\underline{x}) = \frac{y(x_1, \ldots, x_i + \Delta, \ldots, x_k) - y(\underline{x})}{\Delta} \qquad (2.1)$$

where y is the model output, that is in our case, the performance indicator TO, $x_i \leq 1 - \Delta$ and Δ is a predetermined multiple of $1/(p-1)$.

The estimates for the mean and standard deviation of the elementary effects are based on independent random samples of the elementary effects. The samples are obtained by application of carefully constructed sampling plans.

The general procedure to assess one single sample for the elementary effect of each parameter is as follows. Initially, the parameter vector is assigned a random base value (on the discretized grid). An observation of the model output is made. Then a "path" of k orthogonal steps through the k-dimensional parameter space is followed. The order of the steps is randomized. After each step an observation is made and the elementary effect associated with that step is assessed.

With this procedure, a set of r independent samples for the elementary effects can be obtained by repeating this procedure r times. An illustration for a three-dimensional parameter space is presented in Figure 2.4.

This concludes the brief introduction to the principles of uncertainty analysis in this subsection. The next subsections show how these principles are applied in the example case.

2.3.3 Uncertainty in model parameters

As a first step in this crude uncertainty analysis, we will assess plausible ranges for the model parameters, globally expressing the uncertainty in their values. In future

Table 2.1 Categories of uncertain model parameters

Description
Physical properties of materials and components
Space dimensions
Wind reduction factor
Wind pressure coefficients
Discharge coefficients
Convective heat transfer coefficients
Albedo
Distribution of incident solar gain:
fraction lost
fraction via furniture to air
fraction to floor
fraction to remainder of enclosure
Air temperature stratification
Radiant temperature of surrounding buildings
Local outdoor temperature

steps of the analysis, these ranges will be interpreted as central 95% confidence intervals. As mentioned in the introduction of the chapter, the parameter uncertainty may arise from two sources namely, specification uncertainty and modeling uncertainty. The specification uncertainty relates to a lack of information on the exact properties of the building. In the case at hand, this mainly concerns the building geometry and the properties of the various materials and (prefabricated) components.

Modeling uncertainty arises from simplifications and assumptions that have been introduced in the development of the model. As a result, the building model contains several (semi-) empirical parameters for which a range of values can be estimated from the literature. Moreover, the model ignores certain physical phenomena.

Table 2.1 shows the list of parameter categories, which have been considered as uncertain. For the case under study a total of 89 uncertain parameters were identified.

A full investigation of the uncertainties in these parameters can be found in de Wit (2001). Here, we will discuss how uncertainty estimates can be made for three different types of parameters. For each of these three parameter types a different approach is used to accommodate the specific features of the uncertainties involved.

Uncertainty in physical properties of materials and components. As the design process evolves, the specification of materials and (prefabricated) components gradually becomes more detailed, but it rarely reaches a level where the physical properties are precisely known. The associated uncertainty is typically specification uncertainty, arising from variations in properties between manufacturers, between batches or even between products within a batch. These variations can be estimated on the basis of an inventory of product data. In this case, data were used from two previous sensitivity analyses in the field building simulation (Pinney *et al.* (1991) and Jensen (1994)) and the underlying sources for these studies (CIBSE (1986), Clarke *et al.* (1990), Lomas and Bowman (1988)). Additional data were obtained from

ASHRAE (1997), ISSO (1994), and the Polytechnic Almanac (1995). For a few parameters a range was assumed for lack of data.

Apart from the uncertainty ranges for the individual material properties, estimates must be made of the statistical dependencies between these properties. If two properties are dependent, that is, have a strong positive correlation, then high values for one property tend to coincide with high values for the other. If the two properties are independent, however, the value of one property does not change the expectations with respect to the value of the other. In this crude uncertainty analysis we will only distinguish two levels of dependency: completely (positively) correlated or uncorrelated.

To estimate the correlations between the properties of different components and materials, each property x has been considered as the output of the hierarchical model:

$$x = \mu_x + \Delta x_1 + \Delta x_2 + \Delta x_3 \tag{2.2}$$

where μ_x is the general mean over the whole population; Δx_1, the variation between types, which satisfy the description in the design specifications; Δx_2, the variation between production batches within a type; and Δx_3, the variation between individual components within a batch.

It has been assumed that the variation in the material and component properties predominantly arises from the first variation component Δx_1. Hence, complete correlation has been considered between properties of the same name, if they belong to components and materials of the same name. Dependencies between different properties or between unlike components or materials have not been considered.

UNCERTAINTY IN WIND PRESSURE COEFFICIENTS

In our case, the ventilation flows through the building are mainly driven by the local (wind) pressures at the locations of the windows in the façades. These pressures depend on the wind velocity upstream of the building, the position on the building envelope, the building geometry, the wind angle with respect to the orientation of the building, the geometry of the direct environment of the building and the shape of the wind profile. In the simulation, only the wind velocity and wind angle are explicitly taken into account, the effect of all other factors is captured in a single coefficient, the wind pressure coefficient. In fact, this coefficient can be considered as a massively simplified model of the airflow around the building and its environment. It is clear that not specification uncertainty, but modeling uncertainty will be dominant for this coefficient.

Several tools have been developed to assist the assessment of mean wind pressure coefficients on the basis of existing experimental data from prior wind tunnel studies and full-scale measurements. The tools from Allen (1984), Grosso (1992), Grosso et al. (1995), Knoll et al. (1995), and Knoll and Phaff (1996) have been applied to the current case to assess the required wind pressure difference coefficients.[2] The results are shown in Figure 2.5. A more detailed analysis of the wind pressure difference coefficients can be found in Section 2.4.

As a start, we will use the intermodel scatter in Figure 2.5 as an estimate for the uncertainty in the wind pressure difference coefficients in this crude uncertainty analysis.

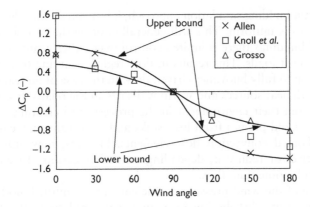

Figure 2.5 Wind pressure difference coefficients from three different models as a function of wind angle (for the definition of the wind angles see Figure 2.2). The figure is symmetric with respect to wind angle 180°, so only the values between 0° and 180° are shown. The drawn lines indicate the upper and lower bounds, which have been used in the uncertainty analysis as central 95% confidence intervals.

These estimates may be inappropriate for several reasons:

- The case at hand is out of the range of application of some of the models. Are the outcomes still appropriate?
- The scatter in the experimental data on which the models are based is eliminated by regression or averaging. Part of this scatter may be measurement error, but part of it results from effects unexplained by the model. Models sharing the same parameters most likely ignore the same effects.
- There is overlap in the data sets underpinning the different models. This overlap introduces a dependency between the model predictions.
- The majority of the data underlying the models that assess the effect of the near field were obtained in (wind tunnel) experiments with regularly arranged near field layouts. The near field in this case is irregular and consists of buildings of different heights.

However, it provides a convenient first estimate for a crude uncertainty analysis. Hence, lower and upper bounds have been used, which are closely tied to the various model results as shown in Figure 2.5. In the analysis, the absolute values of the mean pressure difference coefficients for different wind angles have been considered to be completely and positively correlated. Loosely stated, this means that if the magnitude of the wind pressure difference coefficient for a given wind angle has a high value (relative to its range in Figure 2.5), the pressure differences for all other angles are also large, and vice versa.

Coefficients replacing entire physical models are also used at other common places in simulation models. Examples are the wind reduction factor, heat transfer coefficients and discharge coefficients. Estimates of the uncertainties in these coefficients can be obtained in similar ways as shown here for the wind pressure coefficients.

AIR TEMPERATURE STRATIFICATION

In the mainstream building simulation approach, it is assumed that the air temperature in building spaces is uniform. This will generally not be the case, however. In naturally ventilated buildings there is limited control over either ventilation rates or convective internal heat loads. This results in flow regimes varying from predominantly forced convection to fully buoyancy-driven flow. In the case of buoyancy-driven flow, plumes from both heat sources and warm walls rise in the relatively cool ambient air, entraining air from their environment in the process, and creating a stratified temperature profile. Cold plumes from heat sinks and cool walls may contribute to this stratification. Forced convection flow elements, like jets, may either enhance the stratification effect or reduce it, depending on their location, direction, air stream temperature, and momentum flow.

As in the case of the wind pressure coefficients, the simplified modeling approach of the air temperature distribution in mainstream simulation introduces modeling uncertainty. There is a difference however. Whereas the effect of the airflow around the building on the ventilation flows is reduced to an empirical model with a few coefficients, the effect of temperature stratification in a building space on heat flows and occupant satisfaction is completely ignored. To be able to account for thermal stratification and the uncertainty in its magnitude and effects, we will first have to model it.

If we consider the current approach as a zero-order approximation of the spatial temperature distribution, then it is a logical step to refine the model by incorporating first-order terms. As vertical temperature gradients in a space are commonly dominant, we will use the following model:

$$T_{air}(z) = \overline{T}_{air} + \xi(z - H/2) \tag{2.3}$$

where T_{air} is the air temperature; $\overline{T}_{air}$, the mean air temperature; z, the height above the floor; H, the ceiling height of the space; and ξ, the stratification parameter.

Dropping the assumption of uniform air temperature has the following consequences:

- the temperature of the outgoing air is no longer equal to the mean air temperature as the ventilation openings in the spaces are close to the ceiling;
- the (mean) temperature differences over the air boundary layers at the ceiling and floor, driving the convective heat exchange between the air and those wall components, are no longer equal to the difference between the surface temperature and the mean air temperature;
- the occupants, who are assumed to be sitting while doing their office work, are residing in the lower half of the space and hence experience an air temperature that is different from the mean air temperature.

With Equation (2.3) we can quantify these changes and modify the simulation model to account for them. In most commercially available simulation environments this is not feasible, but in the open simulation toolkit BFEP this can be done.

In the analysis we will assume that ξ in Equation (2.3) is a fixed, but uncertain parameter. This means that we randomize over a wide variety of flow conditions in the space that may occur over the simulated period. In, for examples, Loomans

(1998, full-scale experiments and flow-field calculations) and Chen *et al.* (1992, flow-field calculations) vertical temperature differences over the height of an office space are reported between 0°C and 2°C for mixing ventilation conditions and from 1°C up to 6°C for displacement ventilation configurations. These numbers suggest that vertical temperature differences of several degrees may not be uncommon. Hence, we will choose ξ in the range [0, 1]°C/m in this study.

The temperature stratification effects in separate spaces have been assumed independent. No stratification has been assumed in the corridor between the office spaces in the case at hand.

2.3.4   Uncertainty in scenario

The simulation scenario in this context concerns the outdoor climate conditions and occupation profile. In practice, it has become customary to use standardized scenario elements in comfort performance evaluations. The most striking example concerns the "reference" time series of outdoor climate data. From the experience with performance evaluations, in which these standardized scenario conditions were used, a broad frame of reference has developed to which performance calculations for new buildings can be compared. If such comparisons are indeed meaningful to a decision-maker, who aims to use a performance evaluation to measure the level of achievement of his objectives, there is no scenario uncertainty. If, however, a decision-maker is actually interested in a performance assessment, based on a *prediction* of the comfort sensations of the future occupants of the building, the scenario should be considered as a reflection of the future external conditions, which are uncertain.

As a systematic exploration of a decision-maker's objectives, and their translation into building performances is commonly not undertaken in building design, it is difficult to decide in general how to deal with scenario uncertainty. In this example, we will not address scenario uncertainty.

2.3.5   Propagation of uncertainty

On the basis of the parameter uncertainties identified in the previous section, the uncertainty in the model output, that is, the building performance can be calculated by propagation of the parameter uncertainties through the model.

For lack of explicit information on the parameter distributions, normal distributions were assumed for all parameters from which samples were drawn. The parameter ranges, established in the previous sections, were interpreted as central 95% confidence intervals. Where necessary, the normal distributions were truncated to avoid physically infeasible values.

As discussed in Section 2.3.2.3, the uncertainty in model parameters is propagated by means of Latin Hypercube Sampling, a modified Monte Carlo technique. In this study the algorithm for Latin Hypercube Sampling from UNCSAM (Janssen *et al.* 1992) was applied. A total of 250 samples were propagated, which is well above the value of $4k/3$ ($k = 89$ being the number of parameters) that Iman and Helton (1985) recommend as a minimum.

For each sample of parameter values, a dynamic temperature simulation was carried out. In each simulation, a single, deterministic scenario was used. This

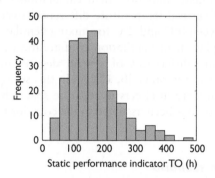

Figure 2.6 Histogram of the performance indicator TO obtained from the propagation of the Latin Hypercube Sample of size 250. Note that uncertainty in the scenario has not been taken into account. A common target value for TO is 150 h.

scenario covers a period of six months from April through September. From the resulting temperature time series, the performance indicator TO was calculated. The results of the propagation of 250 samples are shown in Figure 2.6.

The variability in the comfort performance, observed in the Monte Carlo exercise is significant. For both the static and the adaptive performance indicator the coefficient of variation, that is, the standard deviation divided by the mean value, is about 0.5.

2.3.6   Sensitivity analysis

A parameter screening was carried out with the factorial sampling method according to Morris (1991) as explained in Section 2.3.2.4. The 89 parameters ($k = 89$) were discretized on a 4-level grid ($p = 4$). The elementary step Δ was chosen to be 2/3, as shown in Figure 2.4. For each parameter five independent samples ($r = 5$) of the elementary effects on the comfort performance indicator TO were assessed in 450 simulation runs. The mean value of TO over these runs was 170 h. Figure 2.7 shows for each parameter the sample mean m_d and the standard deviation S_d of the observed elementary effects on the static performance TO.

Important parameters are parameters for which the elementary effect has either a high mean value or a large standard deviation. Table 2.2 shows the five most important parameters found in the screening process in decreasing order of importance.

To explore the importance of interactions and nonlinear effects, the dotted lines, constituting a wedge, are plotted in Figure 2.7. Points on these lines satisfy the equation $m_d = \pm 2\, S_d/\sqrt{r}$, where $S_d/\sqrt{r}$ is the standard deviation of the mean elementary effect. If a parameter has coordinates (m_d, S_d) below the wedge, that is $|m_d| > 2\, S_d/\sqrt{r}$, this is a strong indication that the mean elementary effect of the parameter is nonzero. A location of the parameter coordinates above the wedge indicates that interaction effects with other parameters or nonlinear effects are dominant.

To check if these five parameters indeed account for most of the uncertainty, a Monte Carlo cross-validation was carried out (see Kleijnen 1997; de Wit 2001). This cross-validation showed that the set of five most important parameters explains 85%

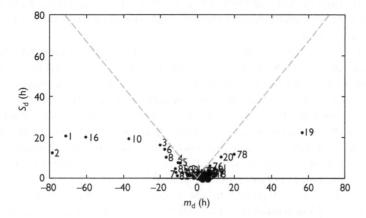

Figure 2.7 Sample mean m_d and standard deviation S_d of the elementary effects on the performance indicator TO obtained in the parameter screening. The numbers in the plot are the parameter indices (see Table 2.2). The dotted lines constituting the wedge are described by $m_d = \pm 2\, S_d/\sqrt{r}$. Points above this wedge indicate significant nonlinear effects or parameter interactions.

Table 2.2 Parameters that emerge from the parameter screening as most important

Index	Description
2	Wind pressure difference coefficients
1	Wind reduction factor
16	Temperature stratification in space under study
19	Local outdoor temperature
10	External heat transfer coefficients

of the total variance, leaving 10% for the remaining 84 parameters and another 5% for interactions. These numbers confirm that the parameters in Table 2.2 are the parameters of interest.

2.3.7 Discussion and conclusions

Three immediate conclusions can be drawn from the results in the previous sections. First, the five parameters in Table 2.2, that is the wind pressure difference coefficients, the wind reduction factor, temperature stratification, local outdoor temperature and the model for the external heat transfer coefficients are the parameters that account for the majority of the uncertainty in the model output.

Second, although several parameters of secondary importance line up along the wedges in Figure 2.7, indicating the presence of parameter interactions or non-linearity of the model output in the parameters, these effects do not seem to play a significant role. Lomas and Eppel (1992) report similar findings in their sensitivity studies on thermal building models. These studies concerned different model outputs (air temperature and plant power) though, and considered a slightly different set of uncertain parameters.

Finally, the variability in the comfort performance assessments, obtained in the Monte Carlo propagation exercise is significant. This is expressed by the coefficient of variation of 0.5 and the histogram in Figure 2.6. In current practice, the simulated value of the performance indicator is commonly compared with a maximum allowed value between 100 and 200 h to evaluate if the design is satisfactory or not under the selected scenario. Figure 2.6 shows that a simulated point value of the performance does not give much basis for such an evaluation. Indeed, simulation results may depict the design as highly satisfactory or as quite the contrary by just changing the values of the model parameters over plausible ranges.

However, the observed spread in the comfort performance values is based on crudely assessed 95% confidence intervals for the model parameters. An improved quantification of the uncertainty in the building performance could be obtained via a more thorough assessment of the parameter uncertainties. Clearly, those parameters that have been ranked as the most important ones deserve primary focus. We will focus on wind pressure coefficients and temperature stratification as they are in the top five and the crude estimates of their uncertainties have been explicitly discussed earlier. The ranges for the most important set of parameters, that is, the wind pressure difference coefficients, have been based on the scatter between various models. Proper use of these models, though, requires wind-engineering expertise, both to provide reliable inputs to the models and to assess the impact of features in the case under study, which are not covered in the models. The uncertainty estimate for the thermal stratification in a space has been based on, hardly more than, the notion that a temperature difference between ceiling and floor of a couple of degrees is not unusual. A fairly crude parameterization of the stratification has been used with an equally crude assumption about the uncertainty in the parameter. As this parameter turns out to be important, the phenomenon deserves further attention, but more merit cannot be attributed to the current uncertainty range or to its contribution to the uncertainty in the building performance.

Summarizing, it is desirable to further investigate the uncertainty in the model parameters, especially the ones identified as most important. The next chapter addresses the uncertainty in both the wind pressure coefficients and the air temperature distribution in more detail.

2.4   Refinement of the uncertainty analysis

2.4.1   Introduction

As discussed in the previous section, the uncertainty estimates of especially the wind pressure coefficients and the thermal stratification deserve primary attention. The uncertainty in those parameters dominates the uncertainty in the building performance and its assessment can be improved in various aspects. In the next two subsections, the uncertainty in these parameters is scrutinized consecutively. Subsequently, it is analyzed to which degree the uncertainty in the building performance, estimated in the initial uncertainty analysis, has to be revised.

2.4.2   Uncertainty in wind pressure coefficients

2.4.2.1   Introduction

To simulate natural ventilation flows in buildings, the wind pressure distribution over the building envelope is required. In the design of low-rise buildings, wind tunnel experiments are scarcely employed to measure these wind pressures. Instead, techniques are used which predominantly rely on inter- or extrapolation of generic knowledge and data, for example, wind pressure coefficients, previously measured in wind tunnel studies and full-scale experiments. Due to the complexity of the underlying physics, this is a process, which may introduce considerable uncertainty.

In the crude uncertainty analysis reported in the previous paragraph, the quantification of this uncertainty did not go beyond the appraisal done by the analyst performing the study. However, the uncertainty in the wind pressure coefficients can more adequately be quantified by experts in the field of wind engineering. These experts are acquainted with the complexity of the underlying physics and hence best suited to interpolate and extrapolate the data they have available on the subject and assess the uncertainties involved. The next section reports on an experiment in which expert judgment was used to quantify the uncertainties in the wind pressure difference coefficients in the case at hand.

2.4.2.2   Principles of an expert judgment study

In an expert judgment study, uncertainty in a variable is considered as an observable quantity. Measurement of this quantity is carried out through the elicitation of experts, namely people with expertise in the field and context to which the variable belongs. These experts are best suited to filter and synthesize the existing body of knowledge and to appreciate the effects of incomplete or even contradictory experimental data. The uncertain variables are presented to the experts as outcomes of (hypothetical)[3] experiments, preferably of a type the experts are familiar with. They are asked to give their assessments for the variables in terms of subjective probabilities, expressing their uncertainty with respect to the outcome of the experiment. Combination of the experts' assessments aims to obtain a joint probability distribution over the variables for a (hypothetical) decision-maker, DM, who could use the result in his/her decision-problem. The resulting distribution, which is referred to as the DM, can be interpreted as a "snapshot" of the state-of-the-knowledge, expressing both what is known and what is not known.

To meet possible objections of a decision-maker to adopt the conclusions of an expert judgment study, which are based on subjective assessments, it is important that a number of basic principles are observed. These include the following:

- *Scrutability/accountability*: all data, including experts' names and assessments, and all processing tools are open to peer review.
- *Fairness*: the experts have no interest in a specific outcome of the study.
- *Neutrality*: the methods of elicitation and processing must not bias the results.
- *Empirical control*: quantitative assessments are subjected to empirical quality controls.

Cooke and Goossens (2000) present a procedure for structured elicitation and processing of expert judgment, which takes proper account of these principles. This procedure was closely followed here. An outline is presented in the following section.

2.4.2.3 Set-up of the experiment

SELECTION OF THE EXPERTS

A pool of candidates for the expert panel was established by screening the literature on relevant issues like wind-induced pressures on low-rise buildings in complex environments and wind-induced ventilation of buildings. From this pool, six experts were selected on the basis of the following criteria:

- access to relevant knowledge;
- recognition in the field;
- impartiality with respect to the outcome of the experiment;
- familiarity with the concepts of uncertainty;
- diversity of background among multiple experts;
- willingness to participate.

QUESTIONNAIRE

The experts were asked to assess the wind pressure difference coefficients for the case at hand. As the wind pressure difference coefficient depends on the wind angle relative to the orientation of the building, they were asked to give their assessments for 12 different wind angles, with intervals of 30° (cf. Figure 2.2). The case was presented to the experts as if it were a hypothetical wind tunnel experiment, as this is a type of experiment the experts were all familiar with.

Each expert's assessment of a coefficient did not consist in a "best estimate", but in a median value plus a central 90% confidence interval expressing his uncertainty. Table 2.3 shows the first part of the table the experts were asked to fill out for each wind angle.

TRAINING OF THE EXPERTS

It would have been unwise to confront the experts with the questionnaire without giving them some training beforehand. None of the experts but one had ever participated in

Table 2.3 Quantile values of the wind pressure difference coefficients to be assessed by the experts for each of the 12 wind angles

Wind angle	Quantile values		
	5%	50%	95%
0°			
30°			
...			

an experiment involving structured elicitation of expert judgment, so they were unacquainted with the motions and underlying concepts of such an experiment. Moreover, acting as an expert entails the assessment of subjective quantile values and subjective probabilities, a task the experts are not familiar with. Extensive psychological research (Kahneman *et al.* 1982; Cooke 1991) has revealed that untrained assessors of subjective probabilities often display severe systematic errors or biases in their assessments.

Hence, a concise training program for the experts was developed (de Wit 1997a), which the experts had to complete before they gave their assessments in the elicitation session.

ELICITATION

In this stage, the core of the experiment, the experts make their judgments available to the analyst. Individual meetings with each expert were arranged. Moreover, the experts were specifically asked not to discuss the experiment among each other. In this way, the diversity of viewpoints would be minimally suppressed.

The elicitation took place in three parts. Prior to the elicitation meeting, each expert prepared his assessments, for example, by looking up relevant literature and making calculations. During the meeting, these assessments were discussed with the analyst, who avoided giving any comments regarding content, but merely pursued clarity, consistency and probabilistic soundness in the expert's reasoning. On the basis of the discussion, the expert revised and completed his assessments if necessary.

Completion of the elicitation coincided with the writing of the rationale, a concise report documenting the reasoning underlying the assessments of the expert. During the writing of this rationale, which was done by the analyst to limit the time expenditure of the expert to a minimum, issues that had not been identified in the meeting were discussed with the expert by correspondence.

COMBINATION OF THE EXPERTS' ASSESSMENTS

To obtain a single distribution for the decision-maker, DM for all pressure coefficients, the experts' assessments must be combined. This involves two steps:

1   Construction of a (marginal) probability distribution from the three elicited quantile values for each variable and each expert.
2   Combination of the resulting experts' distributions for each variable.

Step 1: Construction of probability distributions.   For each variable, three values were elicited from the experts. These values correspond to the 5%, 50%, and 95% quantiles of their subjective probability distribution. Many probability distributions can be constructed, which satisfy these quantiles. The selection of a suitable probability distribution is a technical issue, which is well-covered in Cooke (1991), but falls outside the scope of this chapter.

Step 2: Combination of the experts' distributions.   For each coefficient, a weighted average of the experts' distributions was calculated for use in the uncertainty analysis. The experts' weights were based on their performance, which was obtained from a statistical comparison of their assessments on so-called *seed* variables with measured realizations of these variables. These seed variables were selected such that their assessment required similar knowledge and skills as the assessment of the variables of interest. Moreover, the experts had no knowledge of the measured values. More information can be found in de Wit (2001).

2.4.2.4   Results and discussion

Figure 2.8 shows the assessments of the combined expert. As a reference, the figure also shows measured values of the wind pressure difference coefficients, which were obtained in a separate wind tunnel study that was dedicated to this particular office building. Moreover, two curves are shown, which demarcate the uncertainty intervals (central 95% confidence intervals) used in the crude uncertainty analysis (see Figure 2.8).

   Main questions to be answered are

1   Are the results of the expert judgment study likely as a proper measure of the uncertainties involved?
2   How do the experts' uncertainty assessments compare to the initial uncertainty estimates used in the crude uncertainty analysis in Section 2.3?

Question 1.   This question can be answered on the basis of the seed variables, for which both expert data and measurements are available. Statistical comparison of these two data sets shows how well calibrated the experts are as a measurement instrument for uncertainty. Loosely stated, a well calibrated expert has no bias (tendency to over- or underestimate) and chooses 90% confidence intervals, which are, on the long run, exceeded by the actual values in 10% of the cases.

   In this particular study, we did not need separate seed variables to analyze the experts' performances as measured values of the wind pressure difference coefficients

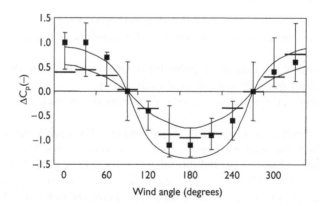

Figure 2.8 Quantile values of the combined expert. The dots are the median values, the error bars represent the central 90% confidence intervals. For reference the wind tunnel results are also shown as horizontal bars. Moreover, the uncertainty estimates used in the crude uncertainty analysis in Section 2.3 are shown as drawn curves.

happened to be available from a separate wind tunnel study that was dedicated to this particular office building.

It can be seen from the figure that all median values of the combined expert are (in absolute value) higher than the measured values. This indicates a bias, that is the experts tend to overestimate the wind pressure coefficients in absolute value. Furthermore, the figure shows that the combined expert's central 90% confidence intervals are exceeded by 1 out of the 12 measured values. Clearly, the experts are well calibrated in this respect.

When both aspects of calibration are combined in one score according to the method of Cooke (1991), it can be concluded that the combined expert is overall fairly calibrated and the results of the expert judgment study are suitable measures of the uncertainty in wind pressure coefficients, which are assessed on the basis of generic wind engineering knowledge and data.

Question 2.   Figure 2.8 shows that overall, the uncertainty assessments from the expert judgment study are somewhat larger than the uncertainty estimates used in the crude uncertainty analysis, especially for the wind angles where the wind approaches over built-up terrain (angles 0–90° and 270–360°). This corroborates the assumption that some sources of uncertainty were omitted in the initial estimates.

The impact of this enlarged uncertainty in the wind pressure coefficients on the building performance is deferred to Section 2.4.4.

2.4.3   Uncertainty in indoor air temperature distribution

In most current simulation tools, the air volume in a building space is typically lumped into one single node, to which a single temperature, that is, the mean air temperature is assigned. Under the assumption that the air temperature is uniform, this air node temperature can be used in the calculation of the ventilation heat flows and the heat flows from the air to the room enclosure on the basis of (semi-) empirical models for the convective heat transfer coefficients. Moreover, the uniform temperature assumption is adopted in the assessment of the average thermal sensation of an occupant in the room.

However, the temperature distribution in the room air will generally not be uniform. Indeed, in naturally ventilated buildings, which are considered in this study, there is limited control over either ventilation rates or convective internal heat loads. This results in flow regimes varying from predominantly forced convection to fully buoyancy-driven flow. In the case of buoyancy-driven flow, plumes from both heat sources and warm walls rise in the relatively cool ambient air, entraining air from their environment in the process, and create a stratified temperature profile. Cold plumes from heat sinks and cool walls may contribute to this stratification. Forced convection flow elements, like jets, may either enhance the stratification effect or reduce it, dependent on their location, direction, temperature, and momentum flow.

In theory, the flow field in a space is fully determined by the Navier–Stokes equations plus the equation for energy conservation with their boundary and initial conditions. When these equations for the flow are solved simultaneously with the other equations in the building simulation model, the two sets of equations supply each other's boundary conditions, and the temperature field is dynamically calculated.

Unfortunately this process is hampered by two problems. First, straightforward solution of the flow equations is not feasible in cases of practical interest. Second, as a result of approximations in the structure and incompleteness of the input of building simulation models, the boundary conditions for the flow are not uniquely specified. This results in uncertainty in the flow field. The results of the sensitivity analysis (see Table 2.3) indicate that this uncertainty gives a potentially significant contribution to the uncertainty in the simulation results. Hence, the aim was to develop an alternative model for the (relevant aspects of the) air temperature distribution, which can readily be integrated in a building model and properly accounts for the uncertainties involved.

An approach was selected that addresses the model development in tandem with an uncertainty analysis. Anticipating significant uncertainty in the air temperature distribution, given the information on boundary conditions in a building simulation context, a coarse heuristic model was proposed with a limited number of empirical parameters. The aim was to assess uncertainty in those parameters and evaluation whether heuristic model and uncertain parameters can suitably describe temperature distribution with its uncertainty.

As for the wind pressure coefficients, expert judgment was used to assess the uncertainties. However, during complications, valid application of expert judgment explicitly requires that the variables which the experts assess are both physically observable and/or meaningful to them. The parameters of the heuristic model did not fulfill this requirement, so an alternative approach was followed.

The experts were asked to assess the main characteristics of the temperature distribution in the space for nine different cases, that is, sets of boundary conditions like wall temperatures, supply flow rates, supply air temperatures, etc. The assessed characteristics, such as mean air temperature and temperature difference over the height of the space, were physically observable. The expert judgment study was set up along the same lines as explained in the previous subsection and resulted in combined uncertainty estimates for all nine cases.

To obtain a (joint) probability distribution over the parameters of the heuristic model, a technique called probabilistic inversion was applied. Probabilistic inversion attempts to find a joint probability distribution over the model parameters such that the model produces uncertainty estimates, which comply with the experts' combined assessments. If the probabilistic inversion is successful, the model plus the resulting joint uncertainty over the model parameters may be taken to properly reflect the air temperature stratification, with its inherent uncertainty, over the range of possible boundary conditions that may occur in the simulations. The probabilistic inversion in this study was carried out with the PREJUDICE-method developed by Kraan (2002). The conclusions from the expert judgment study were very similar to those from the expert judgment study on wind pressures. Again the experts' combined assessments showed a good calibration score, when compared with measured data. The results of the probabilistic inversion showed that a distribution over the 11 model parameters could be found, which reproduced the experts' assessments for 25 out of the 27 elicited variables with sufficient accuracy. The failure of the model to properly reflect the experts' uncertainties on the remaining two variables might, on the basis of the experts' rationales, be attributed to a flaw in the elicitation of the experts. This indicates that the level of detail of the proposed model for the air temperature distribution is well chosen, or more precisely put, not too crude. It is possible that a simpler model would have performed equally well. This could be verified on the basis of the same expert data, as these were collected independently of the model.

Probabilistic inversion has been found to be a powerful tool to quantitatively verify whether the selected level of model refinement is adequate in view of uncertainty in the process, which the model aims to describe. However, it is costly in terms of computation time and in its current form it requires a skilled operator. Hence, the technique is not (yet) suitable in the context of design practice.

2.4.4   Propagation of the uncertainty

The uncertainties that have been identified, augmented by the more refined outcomes of the expert judgment exercises, are propagated through the model to assess the resulting uncertainty in the building performance aspect of interest.

Figure 2.9 shows the results of the propagation of the uncertainty in all parameters. The figures are based on 500 random samples and a fixed scenario (weather data and occupant behavior).

The results in the figure once more confirm that the uncertainty in the indicators for thermal comfort performance is quite pronounced. Compared to the results from the initial crude analysis (Section 2.3), the uncertainty is even somewhat larger. This finds expression in an increase of the coefficient of variation (standard deviation divided by the sample mean) from 0.5 to 0.6. The implications of this uncertainty are the subject of the next section.

An evaluation of this uncertainty on its own merits may give an intuitive idea of its significance and the relevance to account for it in design decision-making. The only way, however, to fully appreciate these issues is by evaluation of the impact of uncertainty information on, or rather its contribution to a design decision analysis.

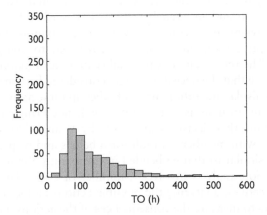

Figure 2.9 Frequency distribution of the comfort performance indicator TO on the basis of 500 samples. The uncertainty in all parameters is propagated.

2.5   Decision-making under uncertainty

2.5.1   Introduction

To ascertain the relevance of uncertainty information, imagine the decision-maker in this case study, who is faced with the choice whether or not to integrate a cooling system in the design of the building case (see Section 2.2). In the particular context, he prefers to implement the cooling system if the TO-performance value of the building (without cooling) will exceed, say, 150 h. To assess the performance, he requests a performance study. The building physics consultant performing the study uses a mainstream simulation approach, which (we hypothesize) happens to turn out a value for TO close to the most likely value according to Figure 2.9, that is 100 h. This value is well below the threshold value of 150 h and the decision-maker comfortably decides not to implement the cooling system. Suppose now that the consultant had not just provided a point estimate, but the full information in Figure 2.9. Then the decision-maker should have concluded that the performance is not at all *well below* the threshold of 150 h. In fact, the probability of getting a building with TO in excess of 150 h is about 1 in 3. In other words, his perception of the decision-problem would have been quite different in the light of the extra information. This in itself is a clear indication that the uncertainty information is relevant for the decision analysis. Hence, the advice should convey this uncertainty in some form.

However, it may not be clear to the decision-maker how to decide in the presence of this extra information. It is no longer sufficient to simply compare the outcome of the performance assessment with a threshold value. To use the information constructively in his decision analysis, the decision-maker needs to weigh his preferences over the possible outcomes (performance values) against the probability of their occurrence. This requires a more sophisticated approach.

2.5.2   Bayesian decision theory

Here, an approach is illustrated, which is based on Bayesian decision theory. Bayesian decision theory is a normative theory; of which a comprehensive introduction and bibliography can be found in French (1993). It describes how a decision-maker *should* decide if he wishes to be consistent with certain axioms encoding rationalism. It is not a prescriptive tool, but rather an instrument to analyze and model the decision-problem. The theory embeds rationality in a set of axioms, ensuring consistency. We will assume that the decision-makers considered here in principle wish their choice behavior to display the rationality embodied in these axioms. If not, a decision analysis on Bayesian grounds is not useful: it will not bring more understanding. Moreover, we assume that decisions are made by a single decision-maker. Choice behavior by groups with members of multiform beliefs and or preferences cannot be rational in a sense similar to that embedded in the axioms alluded to before.

A Bayesian decision analysis involves a number of steps. The first steps include explicating the objectives, analyzing the possible actions and specifying suitable performance indicators to measure the consequences of the actions. For the case at hand, these issues have been discussed in the description of the decision case in Section 2.2. Once these steps have been taken, the consequences of the actions have to be assessed, with their uncertainties, in terms of the performance indicators. This step has been thoroughly discussed in Sections 2.3 and 2.4. To help the decision-maker in making a rational choice between the actions on the basis of this information, decision analysis on the basis of Bayesian decision theory includes a step where the decision-maker explicitly models his preferences.

The crux of Bayesian decision theory is that if a decision-maker adopts the rationality encoded in its underlying axioms, it can be proven that the preferences of the decision-maker can be numerically represented in terms of a function over the performance levels, the *utility* function. In a case that each action leads to a set of attribute levels without uncertainty, the actions can be associated with a single value of the utility function, and the action with the highest utility is preferred. Moreover, if the attribute levels resulting from the actions are uncertain, an action with higher *expected* utility is preferred over one with a lower expected utility. Hence, the *optimal* action is the one with the highest expected utility.

The practical importance of the utility function as a quantitative model for the decision-maker's preference is that it can be assessed by observing the decision-maker's choice behavior in a number of simple reference decision-problems. After this assessment, he can use the function to rank the actions in the actual decision-problem in the order of expected utility. He may directly use this ranking as the basis for his decision or explore the problem further, for example, by doing a sensitivity analysis for assumptions made in the elicitation of either uncertainty or utility, or by a comparison of the expected utility ranking with an intuitive ranking he had made beforehand. Moreover, a systematic assessment of the utility functions helps the decision-maker to clarify and straighten out his own preferences, including the elimination of possible inconsistencies.

2.5.3   Application in the case study

To illustrate the technique, the case described in Section 2.2 will be used. It deals with the situation that only two actions are of concern to the decision-maker, that is, he either leaves the design as it is or he integrates a mechanical cooling system in the design. The two objectives X and Y that are considered are (X) minimizing investment costs and (Y) maximizing occupant satisfaction (measured by the TO) through an investment (cost: 400×10^3 monetary units) in mechanical cooling.

A first step in the actual elicitation of the utility function is the assessment of the (in)dependence structure of this function. The dependence structure indicates in which way the decision-maker's preferences on one attribute depend on the levels of the other attributes. Here we will assume that the decision-maker holds the attributes *additively independent*, which implies that his utility function can be written as

$$U(x, y) = b_2 U_X(x) + b_1 U_Y(y) + b_0 \qquad (2.4)$$

U_X and U_Y are called *marginal* utilities over X and Y. French (1993) briefly addresses the elicitation of (in)dependency structures and gives references. We will not go into that subject here: less strong assumptions about independence lead to similar lines of reasoning as we will follow here although more elaborate. Elicitation of the marginal utility functions U_X and U_Y in a number of simple thought experiments and substitution into (2.4) could result in the decision-maker's utility function (details are given in de Wit (2001)):

$$U(x,y) = -8.3 \times 10^{-4}x - 2.2 \times 10^{-3}y + 1 \tag{2.5}$$

His expected utility is then

$$E\{U(x,y)\} = -8.3 \times 10^{-4}x - 2.2 \times 10^{-3}E\{y\} + 1 \tag{2.6}$$

We used $E\{x\} = x$ here, as the investment cost x is considered to be known without uncertainty. As a result of the linearity of the utility function of this specific decision-maker, we need only limited information on the probability distribution over y, that is only the expected value $E\{y\}$, to calculate the decision-maker's expected utility. We can now calculate the expected utilities for both actions a_1 and a_2 as in Table 2.4. These results suggest that action 1 is the most preferred action of this decision-maker, barring the result of any further analysis the decision-maker might consider.

It is interesting to investigate the result of the analysis for another (imaginary) decision-maker. We assume for the sake of the argument that he differs from decision-maker 1 only in his marginal utility for attribute Y (TO-indicator). Unlike his colleague, he prefers action 1. His line of reasoning might be that buildings with a value of the TO-indicator of 100 h or less are reputedly good buildings with respect to thermal comfort and he is not willing to take much risk that he would end up with a building with TO = 300 h. This decision-maker is *risk averse*. Further elicitation of his marginal utilities might yield the function shown in Figure 2.10.

Table 2.4 Expected utilities for the example
decision-maker

Action	Expected utility
a_1 (zero investment)	0.70
a_2 (maximize comfort)	0.67

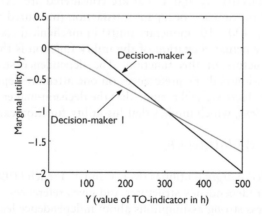

Figure 2.10 Marginal utility function of the two decision-makers over the level of attribute (value of TO-indicator in h).

Table 2.5 Expected utilities for decision-maker 2

Action	Expected utility
a_1 (zero investment)	0.47
a_2 (maximize comfort)	0.50

Following the same approach as for the first decision-maker we arrive at the expected utilities (Table 2.5).

Hence this decision-maker would prefer action 2, whereas his colleague tends to prefer action 1. In itself it is not surprising that two decision-makers with different preferences make different choices in the same situation. However, the two decision-makers in this example would have preferred the same decision in the absence of uncertainty. It is solely as a result of the introduction of uncertainty into the problem that they tend towards different choices.

2.5.3.1 Application in practice

This section discussed how the principles of Bayesian decision analysis can be used as a basis for rational decisions supported by building simulation. Key ingredients of a Bayesian decision analysis are the assessment of the uncertainties in the building simulation results, and explicit modeling of the decision-maker's preferences, for example, in the form of a utility function. In current practice, however, uncertainties in building simulation predictions are not explicitly assessed. Moreover, preference functions in terms of performance are commonly replaced by a set of performance criteria, requiring that each (individual) performance indicator should meet a certain required value.

The gap between the theoretically preferable approach and practical reality is large. Bridging this gap would concern a number of issues. First, a number of technical issues would have to be resolved. A requirement would be the enhancement of the functionality of most building simulation tools to facilitate uncertainty and sensitivity analysis along the lines explained in the previous sections. To use this enhanced functionality effectively, information about uncertainties in model parameters and scenario-elements should be compiled and made available at the fingertips of consultants, who perform building simulation in practical settings.

But the route towards risk-based decision analyses in building practice is hampered by additional barriers. For instance, the costs of building simulation analyses, in terms of time and money, would (significantly) increase. Moreover, consultants and building simulationists would require additional expertise in the fields of statistics, probability theory and decision-making under uncertainty. It is unnerving in this respect that two of the main perceived drawbacks of current, deterministic building simulation are the high level of expertise required to apply building simulation and the high costs related to building simulation efforts (de Wilde and van der Voorden 2003).

These observations suggest that the pervasion of simulation informed, Bayesian decision analysis into building practice doesn't stand a chance. To some extent this suggestion may be correct. Indeed, in many cases a Bayesian decision analysis may point out that the uncertainties in (many of) the performance indicators were not so important to the decision-problem after all. Consider the example in this chapter. If a TO-performance indicator had been found with a mean of 300 h and a standard deviation of 180 h (coefficient of variation equal to 0.6 as in the original example), the decision to implement a cooling system would have been fairly robust under the various possible values of TO. Conversely, if a mean TO-value of 30 h had been found with a standard deviation of 18 h (again the coefficient of variation = 0.6), the option to implement cooling would have been out of the question, (almost) regardless of the precise value of the TO-performance indicator.

This illustrates that it would be beneficial to have a quick scan method, which enables to distinguish the more complex decision-problems from the "clear-cut" cases as sharply as possible. On the one hand, this would make it possible to prevent that a lot of effort is spent on evident cases. On the other hand, it would pinpoint the problems where a more sophisticated analysis would have an added value, justifying extra costs. In designing such a quick scan method, we can learn from the developments in the field of structural reliability analysis, where probabilistic performance evaluation was introduced over 50 years ago (Freudenthal 1947). These developments have resulted in a well-established and well-documented arsenal of methods and tools (Karadeniz and Vrouwenvelder 2003). Although it may not be possible to use all these methods straightforwardly in building simulation-related problems, the methodology behind them is certainly useful.

Probabilistic structural reliability theory is based on the principles of Bayesian decision theory. To translate these principles into tools for mainstream application, three main steps have been taken. The first step is the determination of performance criteria for various performance aspects, a performance criterion being the combination of a performance indicator and a limiting value. The second step is the definition of target probabilities that indicate when a construction does not meet the performance criteria (failure). The third step is the development of verification procedures to check if the performance criteria are met at the required probability levels. In this way, a multi-attribute decision analysis is reduced to one-by-one (probabilistic) verification of individual attributes against fixed requirements.

The question is to which degree this approach would be applicable in the context of typical building simulation informed decision-problems. To analyze this we will discuss the three steps in the approach consecutively.

The first step is the definition of performance limits for the relevant performance indicators. In structural reliability problems related to safety, that is, structural integrity (Ultimate Limit States), the obvious choice for a performance limit is the point of structural collapse. If this limit is exceeded, consequences develop in an almost stepwise manner. In building simulation-related problems, however, there is no such natural choice for a specific performance limit as a decision-maker's preference usually gradually changes with performance (see e.g. Figure 2.10). Hence, performance limits will have a somewhat artificial character.

The second step is the definition of target probabilities for failure. Basically these probabilities should be chosen in such a way that the combinations of failure probability and failure consequences are generally acceptable and do not require unreasonably high investment costs in the majority of the cases. In this step the preferences of an individual decision-maker are replaced by the notions "generally acceptable" and "reasonable." Furthermore, in the original Bayesian decision analysis, the decision-maker could accept performance loss on one aspect if it were sufficiently compensated by a performance gain on another aspect. Such trade-offs cannot be accommodated by performance criteria with fixed target probabilities for individual performance aspects.

So, translation of the first two steps of the standard structural engineering approach to building simulation type problems is not straightforward as it poses quite a number of restrictions on the degrees of freedom of a decision-maker, at least in theory. However, we should realize two things. First, we are investigating the possibilities for a *simplified* method for mainstream application. For those decision-makers' who want the full decision analysis potential at their disposal there is always Bayesian decision theory. Second, most decision-makers in practice are used to these restrictions as the common approach is based on (deterministic) performance criteria. In conclusion, it seems worthwhile to investigate how and to what extent suitable combinations of performance limits and associated target probabilities could be established for building simulation applications along similar lines as in structural reliability.

The third step in the field of structural reliability to make the theory applicable was the development of tools to verify whether the performance criteria are met at the required probability levels. Three types of verification methods were developed, commonly referred to as level I, II, and III methods, respectively.

In level III methods, the probability of failure is calculated fully probabilistically, involving probability distributions over parameters and inputs of the model for performance evaluation. The probability resulting from the calculations can be compared to the target probability. The uncertainty analysis presented in this chapter is level III.

Level II calculations are also fully probabilistic, resulting in an assessment of the failure probability. The difference with level III approaches is that approximations are introduced to speed up the calculations.

Level I calculations are semi-probabilistic calculations based on single values for parameters and inputs called design values. These design values are derived from probabilistic calculations. If the building performance is calculated with these design values and the resulting performance level meets the criterion the probability that the actual performance does not reach the criterion is guaranteed to be less than or equal to the target failure probability.

This level I approach might be a good candidate for the quick scan approach we are looking for. In broad terms, the level I semi-probabilistic calculations would be identical to the common method of verifying performance, but the point estimates for the parameters would be replaced by design values based on probabilistic considerations. Hence, level I building simulations could be carried out without having to apply any probabilistic concepts.

To determine coherent sets of design values, systematic probabilistic analyses are necessary as ad hoc choices may lead to highly uneconomical decisions. An example is mentioned in MacDonald (2002). According to CEN (1998) the declared values of thermophysical data necessary for simulation work are to be quoted for the 90%-fractile. MacDonald estimates that this approach has resulted in plant sizes typically double their necessary size.

2.6   Summary and outlook

This chapter discussed how uncertainty in building simulations can be addressed in a rational way, from a first exploration up to the incorporation of explicit uncertainty information in decision-making. In the context of an example case the structure of an uncertainty analysis was explained, including assessment of the uncertainty in model parameters, propagation of the uncertainty and sensitivity analysis. It was shown how the uncertainty analysis can be specifically refined, based on the results of the sensitivity analysis, using structured expert judgment studies. Finally, this chapter discussed how Bayesian decision theory can be applied to make more rational building simulation informed decisions with explicit uncertainty information.

If explicit appraisal of uncertainty is to pervade building simulation, especially in practical settings, several challenges have to be dealt with:

- *Simulation tools*: the functionality of most building simulation tools needs enhancement to facilitate uncertainty and sensitivity analysis.
- *Databases*: information about uncertainties in model parameters and scenario-elements should be compiled and made available at the fingertips of consultants, who perform building simulation in practical settings.
- *Decision support*: a full Bayesian decision analysis is too laborious for mainstream application. A quick scan method would be indispensable to distinguish the more complex decision-problems from the "clear-cut" cases as sharply as possible.
- *Expertise*: to adequately analyze and use uncertainty information, consultants and building simulationists would require some background in the fields of statistics, probability theory, and decision-making under uncertainty.

This chapter has mainly focused on uncertainty in the context of decision-making. However, the notions and techniques explicated here can also make a contribution in the development and validation of building simulation models. Specific attention can be given to those parts of the model, which give a disproportionate contribution to the uncertainty. If a model part causes too much uncertainty, measures can be considered such as more refined modeling or collection of additional information by, for example, an experiment. On the other hand, model components that prove to be overly sophisticated may be simplified to reduce the time and effort involved in generating model input and running the computer simulations.

It is worthwhile to explore how these ideas could be practically elaborated.

Notes

1 Note that '*simulation*' in Monte Carlo simulation refers to statistical simulation, rather than building simulation.
2 The wind pressure difference coefficient is the difference between the pressure coefficient for the window of modeled building section in the west façade and the one for the window in the east façade.
3 The hypothetical experiments are physically meaningful, though possibly infeasible for practical reasons.

References

Allen, C. (1984). "Wind pressure data requirements for air infiltration calculations." Report AIC-TN-13-84 of the International Energy Agency—Air Infiltration and Ventilation Centre, UK.
Andres, T.H. (1997). "Sampling methods and sensitivity analysis for large parameter sets." *Journal of Statistical Computation and Simulation*, Vol. 57, No. 1–4, pp. 77–110.

ASHRAE (1997). *Handbook of Fundamentals*. American Society of Heating Refrigerating and Air-Conditioning Engineers, Atlanta, Georgia.

Augenbroe, G.L.M. (1982–1988). *BFEP Manual*. Delft University, Delft.

Augenbroe, G.L.M. (1986). "Research-oriented tools for temperature calculations in buildings." In *Proceedings of the 2nd International Conference on System Simulation in Buildings*, Liege.

Augenbroe, Godfried (2000). "The role of ICT tools in building performance analysis." IBPC 2000 Conference, Eindhoven, pp. 37–54.

CEN (1998). "Thermal insulation—building materials and products—determination of declared and design thermal values." Final draft prEN ISO 10456.

Chen, Q., Moser, A., and Suter, P. (1992). "A database for assessing indoor air flow, air quality, and draught risk." Report International Energy Agency, Energy Conservation in Buildings and Community Systems Programme, Annex 20: Air flow patterns within buildings, Subtask 1: Room air and contaminant flow, Swiss Federal Institute of Technology, Zurich.

CIBSE (1986). "Design data." *CIBSE Guide*. Volume A, Chartered Institution of Building Services Engineers, London, UK.

Clarke, J.A. (1985). *Energy Simulation in Building Design*. Adam Hilger, Bristol and Boston.

Clarke, J., Yaneske, P., and Pinney, A. (1990). "The harmonisation of thermal properties of building materials." Report CR59/90 of the Building Research Establishment, Watford, UK.

Cooke, R.M. (1991). *Experts in Uncertainty*. Oxford University Press, New York.

Cooke, R.M. and Goossens, L.H.J. (2000). "EUR 18820—procedures guide for uncertainty analysis using structured expert judgment." Report prepared for the European Commission, ISBN 92-894-0111-7, Office for Official Publications of the European Communities, Luxembourg.

DOE (2003). US Department of Energy Building Energy Software Tools Directory. URL: http://www.eere.energy.gov/buildings/tools_directory/, accessed 28 September 2003.

ESRU (1995). "ESP-r, A building energy simulation environment." User Guide Version 8 Series, ESRU Manual U95/1, Energy Systems Research Unit, University of Strathclyde, Glasgow, UK.

Feustel, H.E. (1990). "Fundamentals of the multizone air flow model COMIS." Report of the International Energy Agency, Air Infiltration and Ventilation Centre, UK.

French, S. (1993) *Decision Theory—An Introduction to the Mathematics of Rationality*, Ellis Horwood, London.

Freudenthal, A.M. (1947). "The safety of structures." *Transactions, ASCE*, Vol. 112, pp. 125–180.

Fürbringer, J.-M. (1994). "Sensibilité de modèles et de mesures en aéraulique du bâtiment à l'aide de plans d'expériences." PhD thesis nr. 1217. École Polytechnique Fédérale de Lausanne, Switzerland.

Grosso, M., Marino, D., and Parisi, E. (1995). "A wind pressure distribution calculation program for multizone airflow models." In *Proceedings of the 3rd International Conference on Building Performance Simulation*, Madison, USA, pp. 105–118.

Iman, R.L. and Conover, W.J. (1980). "Small sample sensitivity analysis techniques for computer models, with application to risk assessment." *Communications in Statistics*, Vol. B11, pp. 311–334.

Iman, R.L. and Helton, J.H. (1985). "A comparison of uncertainty and sensitivity analysis techniques for computer models." Internal report NUREG ICR-3904, SAND 84-1461, Sandia National Laboratories, Albuquerque, New Mexico, US.

ISSO (1994). "Reference points for temperature simulations (in Dutch)." Publication 32, ISSO, Rotterdam, The Netherlands.

Janssen, P.H.M., Slob, W., and Rotmans, J. (1990). "Sensitivity analysis and uncertainty analysis: an inventory of ideas, methods and techniques." Report 958805001 of the Rijksinstituut voor volksgezondheid en milieuhygiene, Bilthoven (RIVM), The Netherlands.

Jensen, S.Ø. (ed.) (1994). "EUR 15115 EN—The PASSYS Project, Validation of building energy simulation programs; a methodology." Research report of the subgroup model validation and development for the Commission of the European Communities. DG XII, Contract JOUE-CT90-0022, Thermal Insulation Laboratory, Technical University of Denmark, Copenhagen, Denmark.

Kahneman, D., Slovic, P., and Tversky, A. (eds.) (1982). *Judgment Under Uncertainty: Heuristics and Biases*, Cambridge University Press, New York.

Karadeniz, H. and Vrouwenvelder, A. (2003). "Overview of reliability methods." SAFEREL-NET Task 5.1 Report SAF-R5-1-TUD-01(5), Saferelnet, Lisboa, Portugal.

Kleijnen, J.P.C. (1997). "Sensitivity analysis and related analyses: a review of some statistical techniques." *Journal of Statistical Computation and Simulation*, Vol. 57, pp. 111–142.

Knoll, B. and Phaff, J.C. (1996). "The C_p-generator, a simple method to assess wind pressures (in Dutch)." *Bouwfysica*, Vol. 7, No. 4, pp. 13–17.

Knoll, B., Phaff, J.C., and Gids, W.F. de (1995). "Pressure simulation program," In *Proceedings of the 16th AIVC Conference on Implementing the Results of Ventilation Research*. Palm Springs, California, USA.

Kraan, B.C.P. (2002). "Probabilistic inversion in uncertainty analysis and related topics." PhD thesis. Delft University of Technology, Delft, The Netherlands.

Liddament, M.W. (1986). "Air infiltration calculation techniques—an applications guide." Report AIC-AG-1-86 of the International Energy Agency—Air Infiltration and Ventilation Centre, UK.

Lomas, K.J. (1993). "Applicability study 1—Executive summary," Report for the Energy Technology Support Unit, contract ETSU S 1213, School of the Built Environment, De Montfort University, Leicester, UK.

Lomas, K.J. and Bowman, N.T. (1988). "Developing and testing tools for empirical validation." In *An Investigation into Analytical and Empirical Validation Techniques for Dynamic Thermal Models of Buildings*, Final BRE/SERC report, Vol. VI, Building Research Establishment, Watford, UK.

Lomas, K.J. and Eppel, H. (1992). "Sensitivity analysis techniques for building thermal simulation problems." *Energy and Buildings*, Vol. 19, pp. 21–44.

Loomans, M.G.L.C. (1998). "The measurement and simulation of indoor air flow." PhD thesis. Eindhoven University of Technology, Eindhoven, The Netherlands.

MacDonald, I.A. (2002). "Quantifying the effects of uncertainty in building simulation." PhD thesis. University of Strathclyde, Scotland.

MacDonald, I.A. (2003). "Applying uncertainty considerations to building energy conservation equations." In *Proceedings Building Simulation 2003, 8th International IBPSA Conference*, August 11–14, Eindhoven, The Netherlands.

MacDonald, I.A., Clarke, J.A., and Strachan, P.A. (1999). "Assessing uncertainty in building simulation." In *Proceedings of Building Simulation, '99*, Paper B-21, Kyoto.

Martin, C. (1993). "Quantifying errors in the predictions of SERI-RES." Report of the Energy Technology Support Unit, contract ETSU S 1368, Energy Monitoring Company Ltd, UK.

McKay, M.D. (1995). "Evaluating prediction uncertainty." Report for the Division of Systems Technology, Office of Nuclear Regulatory Research, USNRC, contract NUREG/CR-6311, Los Alamos National Laboratory, Los Alamos.

McKay, M.D., Beckmann, R.J., and Conover, W.J. (1979). "A comparison of three methods for selecting values of input variables in the analysis of output from a computer code." *Technometrics*, Vol. 21, pp. 239–245.

Morris, M.D. (1991). "Factorial sampling plans for preliminary computational experiments." *Technometrics*, Vol. 33, No. 2, pp. 161–174.

Pinney, A.A., Parand, F., Lomas, K., Bland, B.H., and Eppel, H. (1991). "The choice of uncertainty limits on program input parameters within the applicability study 1 project."

Applicability Study 1. Research report 18 for the Energy Technology Support Unit of the Department of Energy, UK, Contract no E/5A/CON/1213/1784, Environmental Design Unit, School of the Built Environment, Leicester Polytechnic, UK.

Polytechnical Almanac (in Dutch) (1995). *Koninklijke PBNA*, Arnhem, The Netherlands.

Rahni, N., Ramdani, N., Candau, Y., and Dalicieux, P. (1997). "Application of group screening to dynamic building energy simulation models." *Journal of Statistical Computation and Simulation*, Vol. 57, pp. 285–304.

Reedijk, C.I. (2000). "Sensitivity analysis of model output—performance of various local and global sensitivity measures on reliability problems." Report 2000-CON-DYN-R2091 of TNO Building and Construction Research, Delft, The Netherlands.

Saltelli, A., Chan, K., and Scott, E.M. (2000). *Sensitivity Analysis*. Wiley, New York.

Savage, L.J. (1954) *The Foundations of Statistics*. Wiley, New York.

Wijsman, A.J.Th.M. (1994). "Building thermal performance programs: influence of the use of a PAM." BEP '94 Conference, York, UK.

de Wilde, P. and van der Voorden, M. (2003). "Computational support for the selection of energy saving building components." In *Proceedings of Building Simulation 2003, 8th International IBPSA Conference*, August 11–14, Eindhoven, The Netherlands.

de Wit, M.S. (1997a). "Training for experts participating in a structured elicitation of expert judgment—wind engineering experts." Report of B&B group, Faculty of Civil Engineering, Delft University of Technology, Delft.

de Wit, M.S. (1997b). "Influence of modeling uncertainties on the simulation of building thermal comfort performance." In *Proceedings of Building Simulation '97, 5th International IBPSA Conference*, September 8–10, Prague, Czech Republic.

de Wit, M.S. (1997c). "Identification of the important parameters in thermal building simulation models." *Journal of Statistical Computation and Simulation*, Vol. 57, No. 1–4, pp. 305–320.

de Wit, M.S. (2001). "Uncertainty in predictions of thermal comfort in buildings." PhD thesis, Delft University, Delft.

de Wit, M.S. and Augenbroe, G.L.M. (2002). "Analysis of uncertainty in building design evaluations and its implications." *Energy and Buildings*, Vol. 34, pp. 951–958.

Chapter 3

Simulation and uncertainty

Weather predictions

Larry Degelman

3.1 Introduction

Everyone deals with uncertainty every day—whether predicting the outcome of an election, a football game, what the traffic will be like or what the weather will be. Most of us have become accustomed to erroneous predictions by the weather forecasters on television, but we seem willing to accept this sort of uncertainty. No forecaster will give you 100% assurance that it will rain tomorrow; instead, they will only quote to you a probability that it will rain. If it doesn't rain the next day, we usually conclude that we must have been in the "nonprobable" area that didn't receive rain; we don't usually sue the weather forecaster. This type of prediction is done by computerized simulation models, and in fact, these simulation models are not intended to produce one specific answer to a problem. Rather, the underlying premise of simulation is that it discloses a range of situations that are most likely to occur in the real world, not necessary a situation that will definitely occur. This is a very useful aspect to a building designer, so as not to be confined to a single possibility. In short, simulation allows you to cover all the bases.

When we use simulation models to predict thermal loads in buildings, we should recognize that there would be built-in uncertainties due in part to the weather data that we use to drive the simulation models. Most forecasters agree that the best predictor of weather conditions is the historical record of what has occurred in the past. The same forecasters, however, would agree that it is very unlikely that a future sequence of weather will occur in exactly the same way that it did in the past. So, what kind of weather can be used to drive energy simulation models for buildings? what most simulationists would like to have is a pattern of "typical weather"? This entails finding (or deriving) a statistically correct sequence of weather events that typify the local weather, but not simply a single year of weather that has happened in the past.

In this chapter, a simulation methodology is introduced that is intended for application to the climate domain. Featured is the Monte Carlo method for generating hourly weather data, incorporating both deterministic models and stochastic models. Overall, the simulation models described here are targeted toward synthetic generation of weather and solar data for simulating the performance of building thermal loads and annual energy consumption. The objective is not to replace measured weather with synthetic data, for several reliable sources already exist that can provide typical weather data for simulation processes. The modeling methods introduced are also not intended to forecast weather conditions of the type we are accustomed to seeing on television. Rather, the model described here is intended to provide a likely sequence of hourly weather parameters when such data are not available from any measured source. Only statistical parameters need be available. Parameters that are not of particular interest to building thermal loads (such as rain, snow, pollen count, and visibility) will not be addressed. Parameters that are included in the modeling are dry-bulb temperature, humidity (dew-point temperature), solar radiation, wind speed, and barometric pressure.

Specifically, the modeling addresses the following parameters.

Sun–earth variables (daily):
   Solar declination angle
   Variation in the solar constant
   Equation of time
   Time of sunrise and sunset
Solar/site-related data (hourly):
   Sun's altitude and azimuth angles
   Direct normal radiation
   Solar radiation on horizontal surface (direct, diffuse, and total)
Sky data (daily):
   Atmospheric extinction coefficient
   Cloud cover fraction
Temperature data (hourly):
   Dry-bulb
   Dew-point
Relative humidity (from dry-bulb and dew-point temperatures)
Barometric pressure (hourly)
Wind speed (hourly)

Several statistical methodologies have been investigated for generating weather data for thermal simulations in buildings (e.g. Adelard *et al.* 1999). The models and procedures illustrated in this chapter will demonstrate but one approach developed by the author (Degelman 1970, 1976, 1997).

3.2   Benefits of building simulation

Two systems that have strong interrelationships and that affect a major portion of a building's cost are the thermal envelope (roof and walls) and the air-conditioning system. Minimizing the cost of a wall or roof system by cutting back on insulation material is not usually the proper approach to use in minimizing the total cost of a building. If wall and roof constructions are minimal, the heat gains and heat losses will be larger throughout the life of a building, and the operating costs will be larger. Experience shows that it is preferable to use more expensive wall and roof systems to attain better insulation values, which in turn results in a savings over the life cycle of the building.

Simulation of a building's energy performance is a way to help designers calculate life cycle costs and thus optimize the building's ultimate cost and performance. Simulations of this sort are routinely being accomplished every day. The major driving mechanism of thermal heat flows in buildings is the climate. All computer programs require input of extensive weather and solar radiation data, usually on an hourly basis. If these data are readily available, there is no need to simulate the weather; however, when the hourly data are lacking, there is a bonafide need for simulated weather sequences. Even if hourly weather data are available, sometimes it might only represent a few years of record. Such a short record is only anecdotal and cannot purport to represent long-term "typical" weather. The only way to represent the full spectrum of weather conditions that actually exist is to collect data from many years (ten or more) of hourly weather data. Very few weather sites have reliable contiguous weather data available for extremely long periods of time. If they do have the data, it usually has to be reduced to a "typical" year to economize in the computer run time for the simulations. In addition, users can be frustrated over frequent missing weather data points.

This situation can be elegantly addressed by a model that generates hourly weather data for any given location on the earth. There are never any missing data points and reliable predictions can be made of peak thermal load conditions as well as yearly operating costs. This chapter presents such a model. The variables in this simulation technique are kept as basic as possible so that the technique can be applied to estimating heat gains and losses in buildings at any location on earth where scant weather statistics are available. The calculations that establish the actual heat gains and heat losses and the air conditioning loads are not described here, but these methods can be found in other publications (Haberl *et al.* 1995; Huang and Crawley 1996; ASHRAE 2001).

Establishing "typical" weather patterns has long been a challenge to the building simulation community. To this date, there are various models: for example, WYEC (Weather Year for Energy Calculations) (Crow 1983, 1984), TRY (Test Reference Year) (TRY 1976), TMY (Typical Meteorological Year) (TMY 1981), TMY2 (Stoffel 1993; TMY2 1995), CWEC (Canadian Weather for Energy Calculations), and IWEC (International Weather for Energy Calculations). None of these models are based on simulation; rather, they are based on meticulous selections of typical "real weather" months that make up a purported "typical year." These models should be used if they are available for the locale in which the building is being simulated; however, an alternative approach (i.e. synthetic generation) is called for when these weather records are not available.

3.3 The Monte Carlo method

One approach that can be used to simulate weather patterns is use of a random sampling method known as the Monte Carlo method. The Monte Carlo method provides approximate solutions to a variety of mathematical problems by performing statistical sampling experiments on a computer. The method applies to problems with no probabilistic content as well as to those with inherent probabilistic structure. The nature of weather behavior seems to be compatible with this problem domain.

Before the Monte Carlo method got its formal name in 1944, there were a number of isolated instances of similar random sampling methods used to solve problems. As early as the eighteenth century, Georges Buffon (1707–88) created an experiment that would infer the value of PI = 3.1415927. In the nineteenth century, there are accounts of people repeating his experiment, which entailed throwing a needle in a haphazard manner onto a board ruled with parallel straight lines. The value of PI could be estimated from observations of the number of intersections between needle and lines. Accounts of this activity by a cavalry captain and others while recovering from wounds incurred in the American Civil War can be found in a paper entitled "On an experimental determination of PI". The reader is invited to test out a Java implementation of Buffon's method written by Sabri Pllana (University of Vienna's Institute for Software Science) at http://www.geocities.com/CollegePark/Quad/2435/buffon.html.

Later, in 1899, Lord Rayleigh showed that a one-dimensional random walk without absorbing barriers could provide an approximate solution to a parabolic differential equation. In 1931, Kolmogorov showed the relationship between Markov stochastic processes and a certain class of differential equations. In the early part of the twentieth century, British statistical schools were involved with Monte Carlo methods for verification work not having to do with research or discovery.

The name, Monte Carlo, derives from the roulette wheel (effectively, a random number generator) used in Monte Carlo, Monaco. The systematic development of the Monte Carlo method as a scientific problem-solving tool, however, stems from work on the atomic bomb during the Second World War (c.1944). This work was done by nuclear engineers and physicists to predict the diffusion of neutron collisions in fissionable materials to see what fraction of neutrons would travel uninterrupted through different shielding materials. In effect, they were deriving a material's "shielding factor" to incoming radiation effects for life-safety reasons. Since physical experiments of this nature could be very dangerous to humans, they coded various simulation models into software models, and thus used a computer as a surrogate for the physical experiments. For the physicist, this was also less expensive than setting up an experiment, obtaining a neutron source, and taking radiation measurements. In the years since 1944, simulation has been applied to areas of design, urban planning, factory assembly lines and building performance. The modeling method has been found to be quite adaptable to the simulating of the weather parameters that affect the thermal processes in a building. Coupled with other deterministic models, the Monte Carlo method has been found to be useful in predicting annual energy consumption as well as peak thermal load conditions in the building.

The modeling methods described herein for weather data generation include the Monte Carlo method where uncertainties are present, such as day to day cloud cover and wind speeds, but also include deterministic models, such as the equations that describe sun–earth angular relationships. Both models are applied to almost all the weather parameters. Modeling of each weather parameter will be treated in its whole before progressing to the next parameter and in order of impact on a building's thermal performance.

In order of importance to a building's thermal performance, temperature probably ranks first, though solar radiation and humidity are close behind. The next section first describes the simulation model for dry-bulb temperatures and then adds the humidity aspect by describing the dew-point temperature modeling.

3.4 Model for temperatures

3.4.1 Deterministic model

The modeling of temperatures uses both deterministic methods and stochastic methods. The deterministic portion is the shape of the diurnal pattern. This shape is fairly consistent from day to day as shown in Figure 3.1, even though the values of the peaks and valleys will vary.

After the morning low temperature (T_{min}) and the afternoon high temperature (T_{max}) are known, hourly values along the curve can be closely estimated by fitting a sinusoidal curve between the two end-points. Likewise, after the next morning's low temperature (T_{min1}) is known, a second curve can be fit between those two end-points. Derivation of the hourly values then can be done by the following equations.

From sunrise to 3:00 p.m.

$$T_t = T_{ave0} - (\Delta T/2) \cos[\pi(t - t_R)/(15 - t_R)] \tag{3.1}$$

where, T_t is the temperature at time t; T_{ave0}, the average morning temperature, $(T_{min} + T_{max})/2$; ΔT, the diurnal temperature range, $(T_{max} - T_{min})$; π, the universal value of PI = 3.1415927; t_R, the time of sunrise; and 15, the hour of maximum temperature occurrence (used as 3:00 p.m.).

From 3:00 p.m. to midnight

$$T_t = T_{ave1} + (\Delta T'/2) \cos[\pi(t - 15)/(t_{R'} + 9)] \tag{3.2}$$

where T_t, is the temperature at time t; T_{ave1}, the average evening/night temperature, $(T_{max} + T_{min1})/2$; $\Delta T'$, the evening temperature drop, $(T_{max} - T_{min1})$; and $t_{R'}$, the time of sunrise on next day.

From midnight to sunrise the next day

$$T_t = T_{ave1} + (\Delta T'/2) \cos[\pi(t + 9)/(t_{R'} + 9)] \tag{3.3}$$

The time step can be chosen to be any value. Most energy simulation software uses 1-h time steps, but this can be refined to 1-min steps if high precision is required. The thermal time lag of the building mass usually exceeds 1 h, so it is not necessary to use a finer time step than 1 h; however, a finer time step may be desirable when simulating

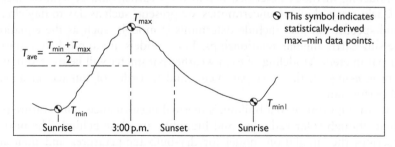

Figure 3.1 Characteristic shape of the daily temperature profile.

on–off timers, daylight sensors, motion sensors, and even thermostats that control the environmental control systems.

That completes the deterministic model for calculating temperatures throughout the course of each day. It should be pointed out that the selection of 3:00 p.m. for the peak daily temperature is merely a norm and may not suit exactly every locality on earth. Also, this value references local standard time, so a 1-h adjustment needs to be made in summers if daylight savings time is utilized.

To calculate hourly values of dew-point temperature, the same procedure can be followed, so there is no need to show additional equations. The main difference between simulation of dew-point temperatures compared to dry-bulb values is that the diurnal curve will be relatively flat, that is, there is little or no rise in the dew-point at 3:00 p.m. Also, one should recognize that the dew-point can never exceed the dry-bulb value at any single point.

3.4.2   Stochastic model

The stochastic portion of the temperature modeling is more intriguing than the deterministic portion, because it has a less prescribed pattern. As a matter of fact, no one knows in advance what the sequence of warm and cold days will be during a month. We see it only after it has happened. It is not actually guesswork, but things are more random than the previous method. This part of the model sets the max–min temperature values for each day and is very much influenced by the uniqueness of the local climate. Fortunately for the simulation community, nature has provided a very well-behaved temperature distribution pattern that nicely fits a bell-shaped curve—better known to the statistical community as the Normal Distribution curve. Statisticians are very familiar with working with Normal distributions. When frequency of occurrences versus the measured variable is plotted, the resulting shape is a bell-shaped curve. This happens when measuring heights of people, areas covered by numerous gallons of paint, or fuel efficiencies attained by a sample of automobiles. Essentially, the highest frequencies of occurrences are around the mean value, while a few are extremely high and a few are extremely low. Average daily temperatures behave in exactly the same way. Furthermore, average daily maximum temperatures also form the same pattern, as do daily minimum temperatures, etc. This distribution pattern is shown in Figure 3.2, depicting the probability of occurrence on the ordinate axis versus the average value plotted on the abscissa.

The probability density function (PDF) for dry-bulb temperatures is almost always Normal (bell-shaped). The mean value is always at the center of the bell, and the spread (fatness) of the bell is determined by the Standard Deviation (σ). As a point of reference, the region bounded between -1σ and $+1\sigma$ always contains 68% of all values. In Figure 3.2, the mean temperature is shown as 72, and the σ is 5.4. The plot shows that the region from -2σ to $+2\sigma$ will contain 95.8% of the temperatures and they will range from 61 to 83. This means that only 4.2% of the values will lie outside this range (half above and half below). Research investigations on temperature occurrences have shown that the "means of annual extremes" (both high and low) are at 2.11σ above and below the mean. This region contains 96.5% of all values. For purposes of weather simulation for energy and design load prediction, it is recommended that this region be utilized. More on this later!

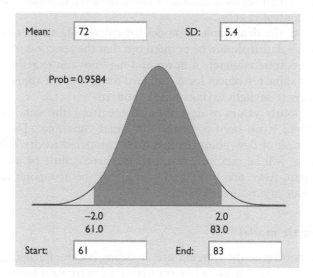

Figure 3.2 The probability density function for the Normal Distribution.

What we would like to do in a simulation application is select daily mean, minimum and maximum temperatures in a manner that would force them to obey the PDF shown in Figure 3.2, that is, that 68 percent of the selections would fall within $\pm 1\sigma$ of the monthly mean and that 96.5 percent of them would fall within $\pm 2.11\sigma$ of the monthly mean. It turns out that the PDF curve is very difficult to utilize for a simulation procedure, so we turn to its integral, the cumulative distribution function (CDF). The CDF is literally the area under the PDF curve starting at the left and progressing toward the right. Its area always progresses gradually from 0 to 1. The literal meaning of the plot is *the probability that a temperature selected from actual measurements will be less than the temperature on the abscissa.* So, we expect lowest temperature from a specific site to be at the left of the graph with probability zero (i.e. the probability is zero that any temperature selection will be less than this minimum value.) Likewise, we expect the highest temperature from a specific site to be at the far right of the graph with probability 1 (i.e. the probability is 100% that any temperature selection will be less than this maximum value). In effect, a CDF plot is made by rank-ordering the temperatures from low to high. Figure 3.3 shows two CDFs with two different standard deviations.

We've played a slight trick in the graph of Figure 3.3. Instead of a probability value showing on the ordinate axis, we show a day of the month. This is simply a method of rescaling of the axis, that is, instead of taking on probability values from 0 to 1, we show days from 1 to 31. This makes the PDF become an instant simulation tool. Say, we randomly pick days in any order, but we select all of them. We first select a day on the ordinate axis, progress horizontally until we intersect the curve, then we progress downward to read the temperature for that day. For convenience, the abscissa is modified to force the mean value to be zero by subtracting all the temperatures from the mean value; thus, the horizontal axis is actually $(T - T_{\text{ave}})$. This

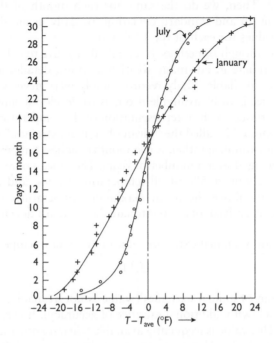

Figure 3.3 Cumulative distribution plots for two separate months.

makes every month normalized to a zero point at the center of its range. By selecting all 31 days and not repeating any day, we can exactly replicate the CDF (and thus the PDF) that had occurred in the actual recorded temperature history of the site. All we need to input to this model is the mean temperature (T_{ave}) and the standard deviation (σ). But, where are standard deviation values obtained?

On a monthly basis, mean values of temperature are readily available for thousands of sites worldwide. Also available are the mean daily maximums and mean daily minimums, and these are frequently available for dew-point temperatures as well. The statistic of Standard Deviation, however, is seldom available in meteorological records. There are two methods available to estimate the standard deviations. The first method is actually to compute it from a long period of records, say 10 or more years. The method is shown by Equation (3.4), as follows:

$$\sigma = \sqrt{\frac{\sum x_i^2 - n\bar{x}^2}{n-1}} \qquad (3.4)$$

where σ is the standard deviation for the period of time studied; n, the number of days in sample; x_i, the daily temperature values (mean or mean maximum values); and $\bar{x}$, the mean temperature for the period studied [$= (\sum x_i)/n$].

It is convenient to use one month for the data collection pool, because most sources of weather data are by published that way. To derive standard deviations for January, for example, we examine historical records for Januaries. If 10 years of data are available, we would examine 310 daily records. The value of n in Equation (3.4) would therefore be 310. Then, we do the same for each month of the year, resulting in a database of a mean and standard deviation for daily mean, daily maximum, and daily minimum values for each of the 12 months.

A less accurate, though sometimes essential, alternative to calculating the standard deviation is to estimate it. For many weather stations, detailed historical records of daily data are not available—daily records simply were never kept. For those sites, another statistic needs to be available if one is to develop a simulation model sufficient enough to produce a close representation of the temperature distributions. The statistic that is required is called the "mean of annual extremes." This is not the average of the daily maximums; rather, it is a month's highest temperature recorded each year and then averaged over a number of years. The "mean of annual extremes" usually incorporates about 96.5% of all temperature values, and this represents 2.11 standard deviations above the mean maximum temperature. Once the "mean of annual extremes" is derived, one can estimate the standard deviation by the equation:

$$\sigma \text{ (est.)} = \frac{\text{mean of annual extremes} - \text{mean maximum temperature}}{2.11} \qquad (3.5)$$

What if the "mean of annual extremes" value is not available? All is not lost—one more option exists (with additional sacrifice of accuracy). It is called "extreme value ever recorded." This value is frequently available when no other data except the mean temperature is available. This is common in remote areas where quality weather recording devices are not available. The "extreme value ever recorded" is approximately 3.1 standard deviations above the mean maximum temperature, so the equation for estimating this becomes

$$\sigma \text{ (est.)} = \frac{\text{extreme value recorded} - \text{mean maximum temperature}}{3.1} \qquad (3.6)$$

As with previous calculations, these standard deviation values need to be calculated for each month of the year.

3.4.3 Random number generation

At this point, we have described how to derive hourly temperatures once minimum and maximum temperatures have been declared. We have seen that the daily average temperatures and daily maximum temperatures are distributed in a Normal Distribution pattern defined by a mean and a standard deviation. Next, we explained that 31 daily values of average temperature could be selected from a cumulative distribution curve (which is also defined by the mean and standard deviation). To select these daily average temperatures, we could simply step through the PDF curve from day 1 through day 31. This would mean the coldest day always occurs on the first day of the month and the hottest day always occurs on the last day of the month, followed by the coldest day of the next month, etc. We realize this is not a realistic representation of weather patterns. So, we need a calculation model to randomly distribute the 31 daily values throughout a month. We might expect to have a few warm days, followed by a few cool days, followed by a few colder days, followed by a few hot days, and finally a few moderate days before the month is over.

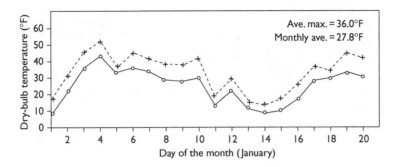

Figure 3.4 Actual record of daily maximum and average temperatures for 20 consecutive days in January.

Figure 3.4 shows a sequence of days from an actual set of recorded month of daily temperature data.

In thermal energy simulations for building applications, we do not necessarily need to replicate this exact sequence; however, we need to replicate the same mean values, the same spread from minimum to maximum, and approximately the same number of "day types" in between the minimum and maximum. Following the CDF in a somewhat random fashion will enable us to meet this objective. So, how do we select the random pattern? It's simpler than it might first appear. In effect, all we have to do is scramble 31 numbers and let each number represent a "day type" on the CDF graph, and thereby derive 31 different temperatures. To randomly order the day types, we use a computerized random number generator. Most computers have inherent random number generators, but there may be reasons why you might want to write your own. The sequence below shows a Fortran code for a random number generator that generates a flat distribution of numbers between 0 and 1, repeating itself only after around 100,000,000 selections. The values derived within the code must have eight significant figures, so double precision variables must be used. Also, an initial value, called "the seed", must be entered to start the sequence. Though we only want 31 numbers, we need to keep internal precision to eight figures, so the sequences won't repeat themselves very often—a lot like the weather. This code can be converted to BASIC with very few modifications.

```
Fortan code for a random number generator
*** RANDOM NUMBER GENERATOR ********************************
    Function RANDOM (B)
    Double Precision B, XL, XK, XNEW
       XL = B * 1E7
       XK = 23.*XL
       XNEW = INT (XK / 1E8)
       B = XK - XNEW * 100000001.
       B = B/1E8
*** Check to be sure B is not outside the range .00001 to 1.
       IF (B.LT.1E-5) B = ABS (B*1000.)
       IF (B.GE.1) B = B/10.
       RANDOM = B
    RETURN
    END
```

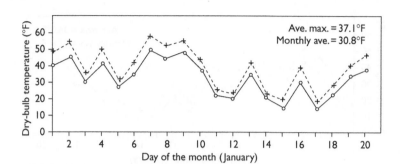

Figure 3.5 Monte Carlo generated daily maximum and average temperatures for 20 consecutive days in January.

Using the random number generator function is similar to "rolling the dice," and is where we finally embrace the concepts on the Monte Carlo method. We start the process by entering a totally meaningless, 8-digit seed value somewhere between 0 and 1 (e.g. 0.29845718). In our software we call the random number generator function by the equation, B = RANDOM(B). The number, B, returned is always an 8-digit number between 0.00001 and 1. Next, we multiply this value by 31 and round up to the next higher integer, creating numbers from 1 to 31. Then, we enter the y-axis of the CDF curve and read the temperature value from the x-axis, the result being the temperature value for that day.

Following this procedure generated the results shown in Figure 3.5 for a selection of the first 20 days. If one were to select a second set of 20 days, a different sequence would result. Every time a series of numbers is selected, a different sequence of days will occur, only repeating the exact sequence after about 100 million trials.

3.4.4 Practical computational methodology

For simplicity in computation of daily temperatures, the means and standard deviations are "normalized" to a Normal Distribution curve with mean $(\mu) = 0$, and standard deviation $(\sigma) = 1$. The 31 possible choices for daily values are shown in Table 3.1. These values range from a low of -2.11 to a high of 2.11 standard deviations, with the center point being 0.

For practical software applications, the CDF values $f(x)$ are stored into a dimensioned array, and the x-value from Table 3.1 is a random variable that only takes on values from 1 to 31. We'll call the dimensioned array FNORMAL(31). The computational sequence is as follows:

(a) Establish X by calling the random number generator and multiplying by 31.

$$X = 31 * RANDOM(rn) \tag{3.7}$$

where rn is the random number between 0 and 1.

Table 3.1 The 31 values of deviations from the mean for a Normal Distribution's Cumulative Distribution Curve

Left half of curve including the mid point

x	1	2	3	4	5	6	7	8	9	10	11	12	13	14	15	16
f(x)	−2.11	−1.70	−1.40	−1.21	−1.06	−0.925	−0.808	−0.70	−0.60	−0.506	−0.415	−0.33	−0.245	−0.162	−0.083	0.0

Right half of curve

x	17	18	19	20	21	22	23	24	25	26	27	28	29	30	31
f(x)	0.083	0.162	0.245	0.33	0.415	0.506	0.60	0.70	0.808	0.925	1.06	1.21	1.40	1.70	2.11

(b) Compute today's average temperature by

$$T_{\text{ave}} = T_{\text{Mave}} + \sigma * \text{FNORMAL}(X) \tag{3.8}$$

where, T_{ave} is the average temperature for today; T_{Mave}, the average temperature for this month; and σ, the standard deviation for average daily temperatures.

Computation of Equations (3.7) and (3.8) is performed 31 times until all the days of the month are completed. The result will be a sequence similar to the pattern shown in Figure 3.5. The pattern will appear to be a bit choppy, so the software developer may wish to apply some biasing to how the 31-day sequence is generated. Usually, there should be 2–4 warm days grouped together before the weather moves to colder or hotter conditions. It is convenient to force the simulation to begin the month at near average conditions and end the month in a similar condition. This prevents large discontinuities when moving from one month to the next, where the mean and standard deviation will take on new values.

If the selection of the days from the cumulative distribution curve is left totally to the random number generator, usually several days will be omitted and several days will be repeated. To obtain the best fit to the normal distribution curve, and thus the best representation of the historical weather, all 31 days should be utilized from the table, and used only once. The methods to do this can be varied. One simple method is to introduce a biased ordering of the day sequence when performing the computer programming. Better conformance to the local climate can be done if the sequence of day selections is correlated to other variables such as solar, humidity, and wind. This requires more extensive analysis of the local climate conditions and may present some rather formidable tasks. This issue will be addressed later in this chapter after the solar simulation techniques have been presented.

3.4.5 Simulation of humidity

The most convenient value to use to represent humidity is the dew-point temperature. It tends to be rather flat during any one day, and its mean value is tightly correlated to the daily minimum temperature. Mean monthly dew-point temperatures are frequently published by the weather stations, but if these are unavailable, they can still be computed from a psychrometric chart assuming that either relative humidity or wet-bulb temperatures are published. One or the other of these is necessary if the dew-point temperature is to be simulated.

The procedure to simulate average daily dew-point temperatures is identical to the dry-bulb temperature simulation method presented in the previous portions of this section. Standard deviations for dew-point are seldom in any publications, so they can simply be set equal to the standard deviations for average daily temperatures. The ultimate control on dew-point temperature has to also be programmed into the software, that is, the dew-point can never exceed the dry-bulb temperature in any one hour or in any one day. This final control usually results in a dew-point simulation that obeys nature's laws and the historical record.

3.5   Model for solar radiation

3.5.1   Introduction

In this section, we will illustrate a model that derives the solar variables. The most significant variables are the sun's position in the sky and the amount of solar radiation impinging on numerous building surfaces, passing through windows, etc. Much of this model can be directly computed by well-known equations. However, since the amount of solar radiation penetrating the earth's atmosphere is dependent on sky conditions, a modeling tool has to be developed to statistically predict cloud cover or other turbidity aspects of the atmosphere. In the latter regard, this model has some similarities to the temperature sequence prediction model in that it follows a stochastic process that is bounded by certain physical laws.

3.5.2   Earth–sun geometry

Predicting solar energy incident on any surface at any time is not difficult if the local sky conditions are known. First, the sun's position is determined by two angles: the altitude angle, β, and the bearing angle, Ψ_z. These angles are shown in Figure 3.6.

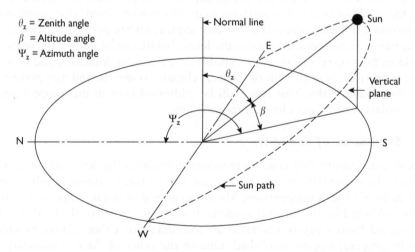

Figure 3.6 Sun position angles.

The altitude angle, β, is measured from the horizontal, and the azimuth angle, Ψ_z, is measured clockwise from North. Those two angles can be computed from the equations that follow

Altitude angle

$$\sin(\beta) = \sin(\delta) * \sin(L) - \cos(\delta) * \cos(L) * \cos(\Gamma) \tag{3.9}$$

where β is the sun altitude angle; δ, the sun declination angle: $\sin(\delta) = \sin(23.5°)$ * $\cos(\pi * D/182.5)$, D, the days measured from June 21st; L, the latitude on earth ($+$N, $-$S); Γ, the hour angle $= \pi * AST/12$, AST $=$ apparent solar time (0–23 h).

Azimuth angle

$$\cos(\Psi_z) = [\sin(L) * \cos(\delta) * \cos(\Gamma) + \cos(L) * \sin(\delta)]/\cos(\beta) \tag{3.10}$$

The apparent solar time (AST) is related to the local standard time (LST) by the equation:

$$AST = LST + ET + 0.15 * (STM - LONG) \tag{3.11}$$

where, AST is the apparent solar time, 0–23 h; LST, the local standard time, 0–23 h; ET, the equation of time, hours; STM, the local standard time meridian; and LONG, the longitude of site measured westward from Greenwich.

The equation of time value can be estimated from a Fourier series representation:

$$ET = -0.1236 \sin(\Omega) + 0.0043 \cos(\Omega) - 0.1538 \sin(2\Omega) - 0.0608 * \cos(2\Omega) \tag{3.12}$$

where ET $=$ equation of time, in hours; $\Omega = \pi$ * (day of the year measured from Jan 1st)/182.5.

3.5.3 Solar radiation prediction

Once the sun position has been determined through use of Equations (3.11) and (3.12), the radiation values can be computed. For visualizing the sun penetration through the atmosphere, we use Figure 3.7. The amount of solar radiation penetrating the earth's

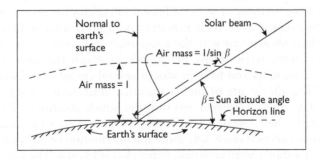

Figure 3.7 Relationship between air mass and altitude angle.

atmosphere is dependent on two factors: the distance the solar beam has to penetrate the atmosphere (known as the air mass) and the degree of sky obscuration (defined by the atmospheric turbidity).

The radiation values are typically segregated into two components—the direct and diffuse. The direct normal insolation utilizes a fairly well-known equation. The generally accepted formula for direct normal solar radiation is

$$I_{DN} = I_o \exp[-a/\sin \beta] \tag{3.13}$$

where I_{DN} is the direct normal insolation; I_o, the apparent solar constant; a, the atmospheric extinction coefficient (turbidity); β, the solar altitude angle; and $1/\sin(\beta)$ is referred to as the "air mass."

The insolation value will be in the same units as the apparent solar constant (usually in W/m^2 or Btu/h per sq. ft.) The apparent solar constant is not truly constant; it actually varies a small amount throughout the year. It varies from around $1,336\,W/m^2$ in June to $1,417\,W/m^2$ in December. This value is independent of your position on earth. A polynomial equation was fit to the values published in *ASHRAE Handbook of Fundamentals* (ASHRAE 2001: chapter 30, table 7):

$$I_o\ (W/m^2) = 1,166.1 + 77.375\ \cos(\Omega) + 2.9086\ \cos^2(\Omega) \tag{3.14}$$

The average value of the apparent solar constant is around 1,167; whereas the average value of the extraterrestrial "true solar constant" is around $1,353\,W/m^2$. This means that the radiation formula (Equation (3.13)) will predicts a maximum of 86% of the insolation will penetrate the atmosphere in the form of direct normal radiation (usually referred to as "beam" radiation).

Everything in Equation (3.13) is deterministic except for a, which takes on a stochastic nature. The larger portion of work is in the establishment of a value for a, the atmospheric extinction coefficient (or turbidity). This variable defines the amount of atmospheric obscuration that the sun's ray has to penetrate. The higher value for a (cloudier/hazier sky), the less the radiation that passes through. ASHRAE publishes monthly values for a, but these are of little value because they are only for clear days. In the simulation process, it is necessary to utilize an infinite number of a values so that the sky conditions can be simulated through a full range of densely cloudy to crystal clear skies.

Fortunately, the methods presented here require that only one value be required to do an hour-by-hour analysis of solar radiation intensities for an entire month. This one value is the average daily solar radiation on a horizontal surface (H). Liu and Jordan (1960) have shown that with the knowledge of this one value, one can predict how many days there were during the month in which the daily solar radiation exceeded certain amounts and have produced several cumulative distribution curves to show this. Through the use of such a statistical distribution, the local sky conditions for each day can be predicted and thus the hourly conditions for each day can also be calculated. Their research results are shown in a set of cumulative distribution curves. These curves (Figure 3.8) show the distribution of daily clearness indices (K_T) when the monthly overall clearness index ($\bar{K}_T$) is known. The Liu–Jordan curves are exceptionally adept for simulation work. In effect, the simulation process works

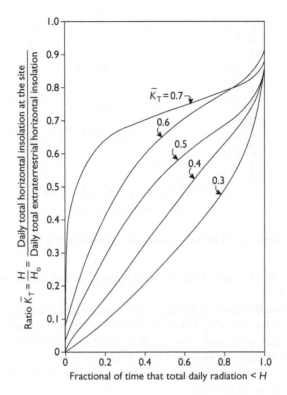

Figure 3.8 The generalized $\bar{K}_T$ curves (from Liu and Jordan 1960).

backward. First, the average daily horizontal insolation (H) is read from the weather station's monthly records. Second, the extraterrestrial horizontal insolation (H_o) is computed outside the atmosphere. The equation for this value is shown here:

$$H_o = (24/\pi)* I_{SC}* [\cos(L)* \cos(\delta)* \sin(SRA) + (\pi - SRA)* \sin(L)* \sin(\delta)] \quad (3.15)$$

where H_o is the extraterrestrial horizontal daily insolation; I_{SC}, the solar constant; SRA, the sunrise angle [$= \pi*$ (sunrise time)/12] measured as compass bearing.

The next step is to derive the monthly $\bar{K}_T$ value by use of the formula:

$$\bar{K}_T = \frac{H}{H_o} \quad\quad\quad (3.16)$$

The $\bar{K}_T$ value derived from Equation (3.16) is then used to determine which monthly $\bar{K}_T$-curve to select from the Liu–Jordan graph in Figure 3.8. The $\bar{K}_T$-curve defines the distribution of daily $\bar{K}_T$ values for all 31 days of a month. The 31 days are evenly distributed along the horizontal axis (between 0 and 1), and for each day a unique $\bar{K}_T$ value is selected. Of course, these days are never entered in a consecutive order; the

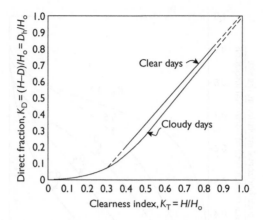

Figure 3.9 Daily horizontal direct fraction versus daily clearness index.

pattern is selected by the same Monte Carlo method described earlier for temperature selections.

In effect, the establishment of the K_T value for any given day is to have established the solar radiation for that day before it really happens. The fourth step is to derive an atmospheric extinction coefficient, a, that will cause the hour-by-hour predictions to add up to this already established value. It is important to have correct sky conditions established, so the breakdown of direct and diffuse radiation components can be done for each hour of the day.

Previous researchers (Liu and Jordan 1960; Perez *et al.* 1990) showed that there is a consistent relationship between daily direct and total global radiation. Their work concludes that both the direct and diffuse portions of solar irradiance can be estimated from the clearness index (K_T) (see Figure 3.9). Because the K_T value is simply the sum of the direct and diffuse portions, equations can be derived for both direct and diffuse fractions. Equations that express these relationships are shown here:

For clear days:

$$K_D = 1.415 * K_T - 0.384 \tag{3.17}$$

For cloudy days:

$$K_D = 1.492 * K_T - 0.492 \qquad \text{for } K_T \geq 0.6, \text{ and} \tag{3.18}$$

$$K_D = \exp(0.935 * K_T^2) - 1.0 \qquad \text{for } K_T < 0.6 \tag{3.19}$$

The K_D value is a weighted average of the sky transmissivity over all the daylight hours. Through examination of a spectrum of cloudy to clear type days, an empirical method has been derived for estimating what the transmissivity for direct radiation would have to be at a known sun angle (say at noon). This work resulted in the formulation of Equation (3.20) for derivation of a, the atmospheric extinction coefficient.

$$a = -\sin(\beta) * \ln[I_{SC} * \tau_D] \qquad (3.20)$$

where τ_D, is the transmissivity of the atmosphere to direct solar.

$$\tau_D = RATIO * K_D \qquad (3.21)$$

where RATIO = empirically derived ratio

$$RATIO = 0.5 + K_T - 0.5\,K_T^2$$

After a is derived, Equation (3.13) should be computed on an hourly basis for all daylight hours. The diffuse component should then be added to the direct portion to obtain the total global irradiance. The diffuse fraction of radiation (K_d) is simply the total fraction (K_T) less the direct fraction (K_D).

$$K_d = K_T - K_D \qquad (3.22)$$

The horizontal diffuse at any given hour is therefore

$$I_{dh} = K_d * I_{SC} * \sin(\beta) \qquad (3.23)$$

The horizontal direct radiation is

$$I_{Dh} = I_{DN} * \sin(\beta) \qquad (3.24)$$

Finally, the total horizontal insolation is

$$I_h = I_{Dh} + I_{dh} \qquad (3.25)$$

3.6   Wind speed simulation

Wind has less impact on building loads that do either temperature, humidity or solar, so it is reasonable to allow for simplification when simulating it. Wind speeds are generally erratic but tend to have a standard deviation which is equal to one-third (0.33) of their average speed. The wind speed model is very simply a selection of non-repeatable daily values from a Normal monthly distribution of average wind speeds. The hourly wind speed is determined by selection of a random number (between 0 and 1) representing a cumulative probability value. When this value is applied to the cumulative distribution curve the hourly value is obtained.

3.7   Barometric pressure simulation

Pressure calculations can be simply based on the elevation above sea level and then varied somewhat during each month based on certain trends that interrelate pressure with solar and temperature patterns. The day-to-day barometric pressure changes do influence the calculation of wet bulb temperatures and relative humidity; however, this influence is relatively small compared to the influences of elevation differences.

3.8   Correlations between the weather variables

Correlation analyses have been applied to weather variables to determine the strength of relationships between temperature, humidity, solar radiation, and wind. Some correlation coefficients as high as 0.85 occur when correlating dry-bulb temperature averages to maximums (which would be expected). Weaker correlations exist between solar radiation and temperature; however, there is at least some positive correlation between the amount of daily solar radiation and the dew-point depression (i.e. the difference between the dry-bulb temperature and dew-point temperature). This is expected since days with high solar values tend to be dryer and hotter, thus depressing the humidity level. Without presenting a precision methodology to deal with these correlations, the topic is only mentioned here to make the programmer aware that there is a need to devise some bias in the randomness process of selecting temperatures that will be compatible with sky conditions that affect solar radiation.

3.9   Some results of Monte Carlo simulations

The following figures show what sort of results one can expect from the simulation model described in this chapter. First, a close-up look at solar radiation for a very clear day and a cloudy day are shown in Figure 3.10. This graph also shows a sample of recorded weather data for days that were similar to those simulated.

What about the ability of the model to predict the proper number of clear and cloudy days which occur during the month? To check the validity of this prediction technique, two graphs were drawn (Figures 3.11 and 3.12) showing the generalized K_T curves from Liu and Jordan, the curve of actual local weather station data, and

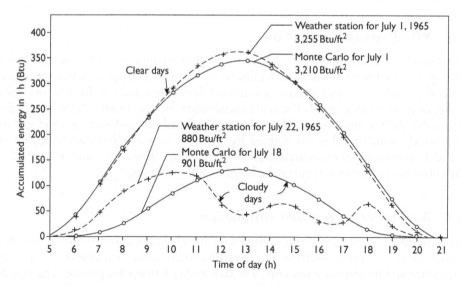

Figure 3.10 Comparison of solar radiation curves from simulated data and actual weather station data for a clear day and a cloudy day.

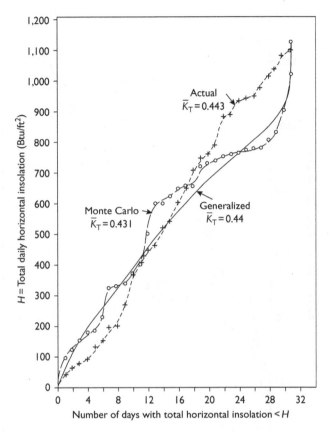

Figure 3.11 Relation between generalized K_T curves, local weather data, and Monte Carlo results for a selected January.

the curve that resulted from a Monte Carlo simulation computer program. Figure 3.11 is drawn for January and Figure 3.12 is for July to show the difference in the shape of the curves. The curve of weather data is for a specific year and does not represent the average for the local weather station readings. The similarity in the characteristic shapes between the actual weather data and the Monte Carlo results indicates a realistic simulation of actual daily weather conditions over the month. The average K_T value of the two curves, however, is likely to be different. This is acceptable since the monthly conditions do, in fact, vary from year to year.

Though statistically generated hourly weather values cannot be compared directly with recorded data, a visible record is always helpful to determine if the model behavior is at least plausible. Figure 3.13 shows a generated sequence of hourly temperatures for one week in April for the city of Bourges, France. Figure 3.14 shows a generated sequence of hourly solar radiation values for a week with clear, overcast and partly cloudy skies for the same city. These illustrate the behavior of the model, though validation is probably better left to comparison of cumulative statistics as will be demonstrated later.

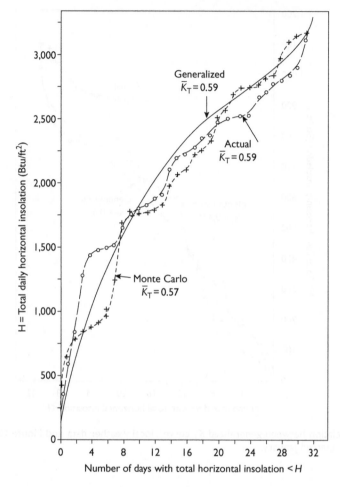

Figure 3.12 Relation between generalized K_T curves, local weather data, and Monte Carlo results for a selected July.

3.10   Validating the simulation model

It is literally impossible to validate a statistical weather generation model on an hourly basis, since there is no way (nor is there intent) to mimic real weather on an hourly basis. In place of this, hourly values are generated by the model, and statistical results must be compared to long periods of recorded weather data statistics. We can demonstrate reliable behavior of the model, for example, by comparing monthly means, standard deviations, and cumulative degree-days between the model and long-term weather records. When this particular simulation model is run, summary statistics are reported at the end of each month; these show both input and output values for means and standard deviations for temperatures and means for solar radiation, wind and barometric pressure. Also tabulated is the difference between the input and output, so the user has an instant reference as to how the means and extremes compare.

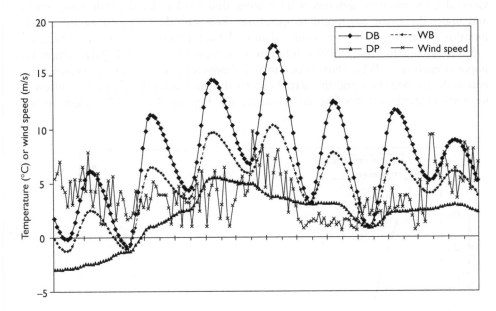

Figure 3.13 Hourly temperatures and wind speeds generated for one April week in Bourges, France.

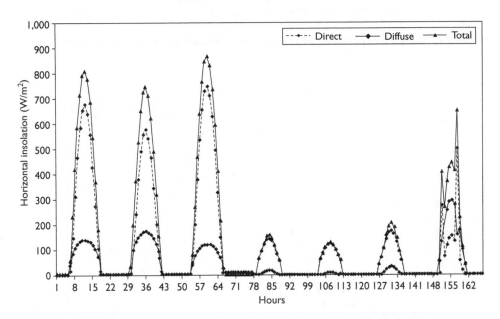

Figure 3.14 Hourly insolation values generated for one April week in Bourges, France, with clear, overcast and partly cloudy skies.

Generally, the monthly differences have fallen under 0.1°C for dry-bulb temperatures and under 0.2°C for dew-point temperatures. For the other parameters, the average monthly difference is 0.9% for wind and around 2% for daily horizontal solar radiation.

Some other efforts at model validation were reported in an ASHRAE symposium paper (Degelman 1981). That work showed almost a perfect agreement between generated degree-day data and the actual recorded degree-days (less than 1% difference in both heating and cooling degree-days, see Figures 3.15–3.17). That analysis

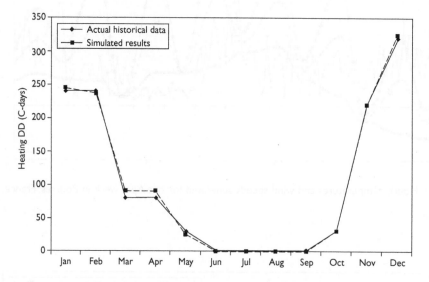

Figure 3.15 Comparison of heating degree-days from simulated versus real weather data.

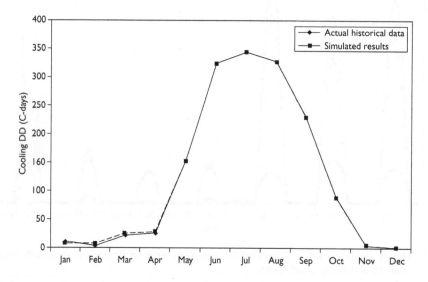

Figure 3.16 Comparison of cooling degree-days from simulated versus real weather data.

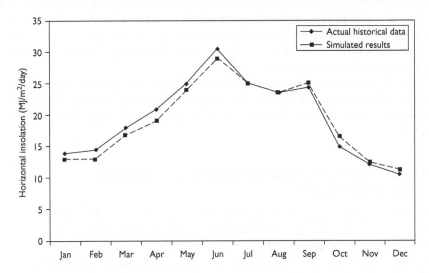

Figure 3.17 Comparison of horizontal daily solar radiation from simulated versus real data.

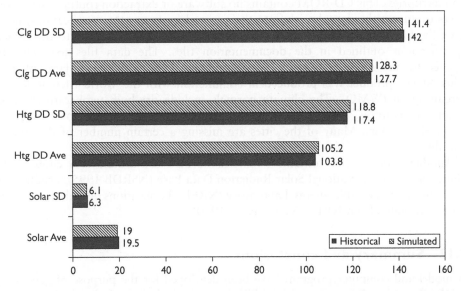

Figure 3.18 Comparison of results from the simulation model to historical weather data. Values represent monthly mean (Ave) and standard deviation (SD) results for heating degree-days, cooling degree-days, and horizontal insolation.

also showed less than a 2% difference in an office building's annual energy consumption when driven by simulated weather data versus the actual weather data from a SOLMET data file of recorded data. The data in Figures 3.15–3.17 are plots of monthly degree-days and solar radiation as derived from the SOLMET file for Dallas, TX, compared to the simulated weather data for the same period. It should be noted that the statistics used in the weather data simulator were actually derived from the same SOLMET data file. This indicates that the simulator is able to re-create at least the dry-bulb temperature and solar distribution with a high degree of accuracy when historical weather data statistics are available.

Figure 3.18 compares the cumulative values from synthetically generated data to actual weather records for Dallas, TX. Shown are average monthly means and standard deviations for heating degree-days, cooling degree-days, and horizontal insolation values.

3.11   Finding input data for driving a statistical weather model

Synthetically generating hourly weather data requires not only a reliable modeling tool, but also a good source of recorded weather statistics. One source of worldwide weather data is available from the National Climatic Data Center (NCDC) in Asheville, NC, USA. On-line access to their publications is possible on their internet site: http://www.ncdc.noaa.gov/oa/climate/climateproducts.html. One product (for sale) is a CD-ROM that contains the 1961–1990 global standard climate normals for over 4,000 stations worldwide, representing more than 135 countries and territories. This CD-ROM contains no software or extraction routines that allow users to import the data directly into their spreadsheets or other applications; however, the files can be read by software written by the user according to the format specifications outlined in the documentation files. The data files may also be opened by any ASCII-compatible application that can handle large data volumes. This NCDC product was produced in conjunction with the World Meteorological Organization (WMO). The climate normals include dry-bulb temperatures, dewpoint temperatures, wind speeds, pressure, and global horizontal solar radiation or sunshine hours. Many of the cities are missing a certain number of the climate variables.

Another product that contains long-term normals for around 248 cities in the United States is the National Solar Radiation Data Base (NSRDB 1995), produced at the National Renewable Energy Laboratory (NREL). Publications describing this data set are available from NREL (Knapp *et al.* 1980).

3.12   Summary and conclusions

A model and computer program have been developed for the purpose of generating synthetic weather data for input to building energy calculation software and sometimes as a replacement for real weather records when real data are hard to find (or are not available). The model has been shown to reliably simulate the variables of temperature, humidity, wind, and solar radiation—all important parameters in computing building beating and cooling loads.

The model testing has been carried out on the basic weather statistics and has been found to be an acceptable representation of real data for the parameters normally regarded to be important to building thermal analyses.

References

Adelard, L., Thierry, M., Boyer, H., and Gatina, J.C. (1999)."Elaboration of a new tool for weather data sequences generation." In *Proceedings of Building Simulation '99*, Vol. 2, pp. 861–868.

ASHRAE (2001). *Handbook of Fundamentals—2001*. American Society of Heating, Refrigerating and Air-Conditioning Engineers, Inc., Atlanta, GA, Chap. 30, p. 30.13.

Crow, Loren W. (1983). "Development of hourly data for weather year for energy calculations (WYEC), including solar data for 24 stations throughout the United States and five stations in southern Canada." Report LWC #281, ASHRAE Research Project 364-RP, Loren W. Crow Consultants, Inc., Denver, CO, November.

Crow, Loren W. (1984). "Weather year for energy calculations." *ASHRAE Journal*, Vol. 26, No. 6, pp. 42–47.

Degelman, L.O. (1970). "Monte Carlo simulation of solar radiation and dry-bulb temperatures for air conditioning purposes." In *Proceedings of the Kentucky Workshop on Computer Applications to Environmental Design*. Lexington, KY, April, pp. 213–223.

Degelman, L.O. (1976). "A weather simulation model for annual energy analysis in buildings." *ASHRAE Trans.*, Vol. 82, Part 2, 15, pp. 435–447.

Degelman, L.O. (1981). "Energy calculation sensitivity to simulated weather data compression." *ASHRAE Trans*, Vol. 87, Part 1, January, pp. 907–922.

Degelman, L.O. (1990). "ENERCALC: a weather and building energy simulation model using fast hour-by-hour algorithms." In *Proceedings of 4th National Conference on Microcomputer Applications in Energy*. University of Arizona, Tucson, AZ, April.

Degelman, L.O. (1997). "Examination of the concept of using 'typical-week' weather data for simulation of annualized energy use in buildings." In *Proceedings of Building Simulation '97, Vol. II, International Building Performance Simulation Association (IBPSA)*. 8–10 September, pp. 277–284.

Haberl, J., Bronson, D., and O'Neal, D. (1995). "Impact of using measured weather data vs. TMY weather data in a DOE-2 simulation." *ASHRAE Trans.*, Vol. 105, Part 2, June, pp. 558–576.

Huang, J. and Crawley, D. (1996). "Does it matter which weather data you use in energy simulations." In *Proceedings of 1996 ACEEE Summer Study*. Vol. 4, pp. 4.183–4.192.

Knapp, Connie L., Stoffel, Thomas L., and Whitaker, Stephen D. (1980). "Insolation data manual: long-term monthly averages of solar radiation, temperature, degree-days and global K_T for 248 National Weather Service Stations," SERI, SP-755-789, Solar Energy Research Institute, Golden, CO, 282 pp.

Liu, Benjamin Y.M. and Jordan, Richard C. (1960). "The interrelationship and characteristic distribution of direct, diffuse and total solar radiation," *Solar Energy*, Vol. IV, No. 3, pp. 1–13.

NSRDB (1995). "Final Technical Report—National Solar Radiation Data Base (1961–1990)," NSRB-Volume 2, National Renewable Energy Laboratory, Golden, CO, January, 290 pp.

Perez, R., Ineichen, P., Seals, R., and Zelenka, A. (1990). "Making full use of the clearness index for parameterizing hourly insolation conditions." *Solar Energy*, Vol. 45, No. 2, pp. 111–114.

Perez, Ineichen P., Maxwell, E., Seals, F., and Zelenda, A. (1991). "Dynamic models for hourly global-to-direct irradiance conversion." In *Proceedings of the 1991 Solar World Congress*. International Solar Energy Society, Vol. I, Part II, pp. 951–956.

Stoffel, T.L. (1993). "Production of the weather year for energy calculations version 2 (WYEC2)." NREL TP-463-20819, National Renewable Laboratory.

Threlkeld, J.L. (1962). "Solar irradiation of surfaces on clear days." *ASHRAE Journal*, Nov., pp. 43–54.

TRY (1976). "Tape reference manual—Test Reference Year (TRY)." Tape deck 9706, Federal
    Energy Administration (FEA), ASHRAE, National Bureau of Standards (NBS), and the
    National Oceanic and Atmospheric Administration (NOAA), National Climatic Center,
    Asheville, NC, September.
TMY (1981). "Typical Meteorological Year (TMY) User's Manual." TD-9734, National
    Climatic Center, Asheville, NC, May, 57 pp.
TMY2 (1995). "User's Manual for TMY2s—Typical Meteorological Years." NREL/SP-463-
    7668, National Renewable Energy Laboratory, Golden, CO, June, 47 pp.

Chapter 4

Integrated building airflow simulation

Jan Hensen

4.1 Introduction

Knowledge of airflow in and around buildings is necessary for heat and mass transfer analysis such as load and energy calculations, for thermal comfort assessment, for indoor air quality studies, for system control analysis, for contaminant dispersal prediction, etc. While airflow is thus an important aspect of building performance simulation, its analysis has considerably lagged behind the modeling of other building features. The main reasons for this seem to be the lack of model data and computational difficulties.

This chapter provides a broad overview of the range of building airflow prediction methods. No single method is universally appropriate. Therefore it is essential to understand the purpose, advantages, disadvantages, and range of applicability of each type of method. The mass balance network modeling approach, and how this is coupled to the thermal building model, is described in more detail. The chapter advocates that the essential ingredients for quality assurance are domain knowledge, ability to select the appropriate level of extent, complexity and time and space resolution levels, calibration and validation, and a correct performance assessment methodology. Directions for future work are indicated.

As indicated in Figure 4.1, building simulation uses various airflow modeling approaches. In terms of level of resolution these can be categorized from macroscopic to microscopic. Macroscopic approaches consider the whole of building, systems, and indoor and outdoor environment over extended periods, while microscopic approaches use much smaller spatial and time scales.

What follows is a brief overview of building airflow modeling methods categorized as semi-empirical or simplified, zonal or computational fluid dynamic modeling approaches. More elaborate descriptions are available in literature (Liddament 1986; Etheridge and Sandberg 1996; Allard 1998; Orme 1999).

4.1.1 *Semi-empirical and simplified models*

These methods are mostly used to estimate air change rate and are frequently based on estimates of building airtightness. A common approach is to estimate the seasonal average air change rate from the building airtightness as measured in a pressurization test. For example by

$$Q = \frac{Q_{50}}{K} \text{ air changes/hour (ach)} \tag{4.1}$$

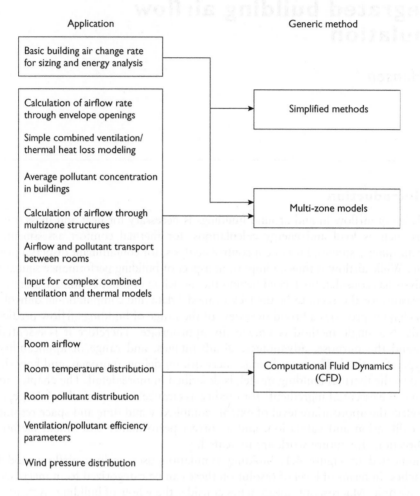

Figure 4.1 Summary overview of typical building airflow applications and modeling techniques.

where Q_{50} is the air change rate at 50 Pa; and K, the empirical constant with value $10 < K < 30$ depending on shielding and characteristics of the building (see Liddament 1986). Often a value of $K = 20$ is applied.

In this example there is effectively no relation with the driving forces of wind and temperature difference. It is possible to improve this with a more theoretically based simplified approach in which Q_{50} leakage data is converted to an equivalent leakage area. The airflow rate due to infiltration is then given by

$$Q = L(A\Delta t + Bv^2)^{0.5} \tag{4.2}$$

where Q, is the airflow rate (L/s); L, the effective leakage area (cm^2); A, the "stack" coefficient; Δt, the average outside/inside temperature difference (K); B, the wind coefficient; and v, the average wind speed, measured at a local weather station.

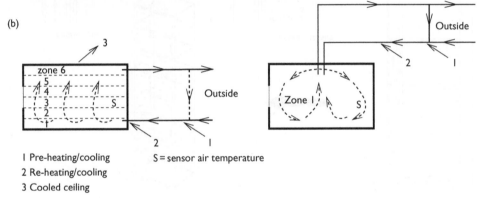

1 Pre-heating/cooling   S = sensor air temperature
2 Re-heating/cooling
3 Cooled ceiling

Figure 4.2 Glasgow's Peoples Palace museum with corresponding simplified airflow model based on rules of thumb; numbers indicate average air change rate per hour (Hensen 1994). (See Plate 1.)

In these types of approaches airflow is modeled conceptually. Based on rules of thumb, engineering values and/or empirical relationships as exemplified earlier, it is up to the user to define direction and magnitude of airflows. A typical application example is shown in Figure 4.2.

In everyday building performance simulation, it is these types of approach that are most commonly used. The main reasons are that they are easy to set up, they are readily understood because they originate from "traditional" engineering practice, and they can easily be integrated with thermal network solvers in building performance simulation software.

4.1.2 *Zonal models*

In a zonal method, the building and systems are treated as a collection of nodes representing rooms, parts of rooms and system components, with internodal connections representing the distributed flow paths associated with cracks, doors, ducts, and the like. The assumption is made that there is a simple, nonlinear relationship between

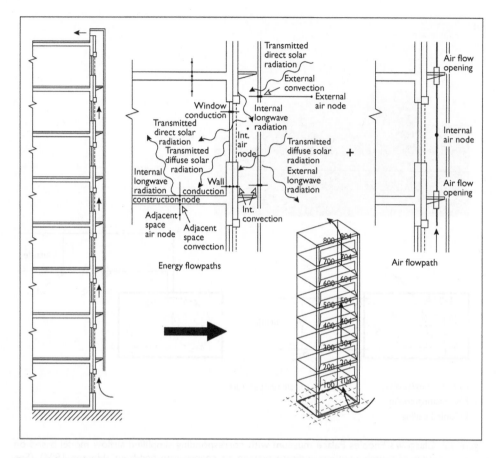

Figure 4.3 Model of a double-skin façade, basically consisting of superimposed coupled thermal and airflow network models (Hensen *et al.* 2002).

the flow through a connection and the pressure difference across it. Conservation of mass for the flows into and out of each node leads to a set of simultaneous, nonlinear equations that can be integrated over time to characterize the flow domain.

Figure 4.3 shows an application example related to prediction of the performance of a double-skin façade system and the impact for the adjacent offices (Hensen *et al.* 2002). In this case the thermal side of the problem is very important. Given the extent of the model and the issues involved, this can only be predicted with building energy simulation. Integration of the network method with building energy simulation is a mature technology (Hensen 1991, 1999a) and nowadays commonly used in practice.

4.1.3 Computational Fluid Dynamics (CFD)

In the CFD approach, the conservation equations for mass, momentum, and thermal energy are solved for all nodes of a two- or three-dimensional grid inside and/or around the building. In structure, these equations are identical but each one represents a different physical state variable. The generalized form of the conservation equation is given by

$$\rho\frac{\partial\phi}{\partial t} + \rho\mu_j\frac{\partial\phi}{\partial k_j} = \frac{\partial}{\partial k_j}\left(\Gamma\frac{\partial\phi}{\partial k_j}\right) + S \tag{4.3}$$

unsteady term + convection term = diffusion term + source term

CFD is a technology that is still very much under development. For example, several different CFD solution methods are being researched for building airflow simulation: direct numerical simulation, large eddy simulation (Jiang and Chen 2001), Reynolds averaged Navier–Stokes modeling, and lattice Boltzmann methods (Crouse *et al.* 2002). In practice, and in the building physics domain in particular, there are several problematic CFD issues, of which the amount of necessary computing power, the nature of the flow fields and the assessment of the complex, occupant-dependent boundary conditions are the most problematic (Chen 1997). This has often led to CFD applications being restricted to steady-state cases or very short simulation periods (Haghighat *et al.* 1992; Martin 1999; Chen and Srebic 2000). An application example is shown in Figure 4.4.

Integration of CFD with building energy is also still very much in development although enormous progress has been made in recent times (Bartak *et al.* 2002; Zhai *et al.* 2002).

Hensen *et al.* (1996) analyzes the capabilities and applicability of the various approaches in the context of a displacement ventilation system. One of the main conclusions of this work is that a higher resolution approach does not necessarily cover all the design questions that may be answered by a lower resolution approach. Each approach has its own merits and drawbacks. An environmental engineer typically needs each approach but at different times during the design process. The main conclusion of this study is summarized in Table 4.1.

Notwithstanding the above, in the context of combined heat and airflow simulation in buildings, it is the zonal method that is currently most widely used. The reasons for this are threefold. First, there is a strong relationship between the nodal networks that represent the airflow regime and the corresponding networks that represent its

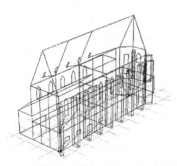 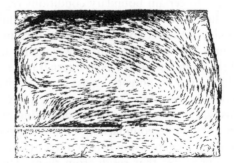

Figure 4.4 Model of a historical building and CFD predictions of air velocity distribution in the central longitudinal section at a particular point in time (Bartak *et al.* 2001).

Table 4.1 Summary of prediction potential ($--$ = none, $++$ = very good) for airflow modeling levels in the context of displacement ventilation system

Aspect	A	B	C
Cooling electricity	$--$	$++$	$--$
Fan capacity	$++$	$++$	$--$
Whole body thermal comfort	$+$	$++$	$+$
Local discomfort, gradient	$--$	$+$	$++$
Local discomfort, turbulence intensity	$--$	$--$	$++$
Ventilation efficiency	$--$	0	$++$
Contaminant distribution	$-$	$-$	$++$
Whole building integration	$++$	$++$	$--$
Integration over time	$++$	$++$	$--$

Note
A = fully mixed zones; B = zonal method; C = CFD (Hensen *et al.* 1996).

thermal counterpart. This means that the information demands of the energy conservation formulations can be directly satisfied. Second, the technique can be readily applied to combined multi-zone buildings and multi-component, multi-network plant systems. Finally, the number of nodes involved will be considerably smaller than that required in a CFD approach and so the additional CPU burden is minimized. The remainder of this chapter will focus on the zonal method.

4.2   Zonal modeling of building airflow

This approach is known under different names such as zonal approach, mass balance network, nodal network, etc., and has successfully been implemented in several software packages such as CONTAMW, COMIS, and ESP-r. The method is not limited to building airflow but can also be used for other building-related fluid flow phenomena such as flow of water in the heating system, etc.

In this approach, during each simulation time step, the problem is constrained to the steady flow (possibly bidirectional) of an incompressible fluid along the connections which represent the building and plant mass flow paths network when subjected to certain boundary conditions regarding pressure and/or flow. The problem reduces therefore to the calculation of fluid flow through these connections with the nodes of the network representing certain pressures. This is achieved by an iterative mass balance approach in which the unknown nodal pressures are adjusted until the mass residual of each internal node satisfies some user-specified criterion.

Information on potential mass flows is given by a user in terms of node descriptions, fluid types, flow component types, interconnections, and boundary conditions. In this way a nodal network of connecting resistances is constructed. This may then be attached, at its boundaries, to known pressures or to pressure coefficient sets that represent the relationship between free-stream wind vectors and the building external surface pressures that result from them. The flow network may consist of several decoupled subnetworks and is not restricted to one type of fluid. However, all nodes and components within a subnetwork must relate to the same fluid type.

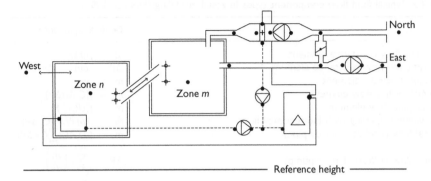

Figure 4.5 Example of building and plant schematic.

Nodes may represent rooms, parts of rooms, plant components, connection points in a duct or in a pipe, ambient conditions and so on. Fluid flow components correspond to discrete fluid flow passages such as doorways, construction cracks, ducts, pipes, fans, pumps, etc. As an example Figure 4.5 shows a schematic of part of a building consisting of two rooms, airflow connections between these rooms, a radiator heating system connected to one zone and an air heating system connected to the other zone. In this case the building and plant configuration contains two mass flow networks—one for air and one for water. One possibility with respect to the translation of this configuration into a fluid flow nodal scheme is indicated by the dots.

In the program, nodes are characterized by several data items, including an identifier, the fluid type, the node type, the height above some arbitrary datum, temperature and several supplementary parameters that depend on the node type. The nodes of the network represent either internal or boundary pressures with only internal nodes being subjected to mass balance tracking. Note that in the present context "internal" is not necessarily equivalent to "inside" nor does "boundary" necessarily equate to "outside". Usually the pressure at an internal node is unknown, although it may be treated as a known parameter as could be required, for example, in the case of an expansion vessel in a hydronic radiator system.

Flow components are characterized by an identifier, a type code (indicating duct, pipe, pump, crack, doorway, etc.) and a number of supplementary data items defining the parameters associated with a specific component type. When a certain flow component is repetitively present in the network, it need only be defined once. Typically supported fluid flow component types are summarized in Table 4.2. Normally each flow component has a subroutine counterpart that is used to generate the flow and flow derivative at each iteration. As an example, the power law component type is elaborated in the next section. Detailed information on other component types can be found elsewhere (Hensen 1990).

A flow network is defined by connections. Each connection is described in terms of the name of the node on its (arbitrarily declared) positive side, the height of the positive linkage point relative to the node on the positive side, the name of the node on the (arbitrarily declared) negative side of the connection, the height of the negative linkage point relative to the node on the negative side, the name of the connecting flow

Table 4.2 Typical fluid flow component types in zonal modeling (Hensen 1991)

Type	General equations
Power law flow resistance element	$\dot{m} = \rho a \Delta P^b$
Quadratic law flow resistance element	$\Delta P = a\dot{m} + b\dot{m}^2$
Constant flow rate element	$\dot{m} = \rho a$
Common orifice flow element	$\dot{m} = f(C_d A \rho \Delta P)$
Laminar pipe flow element	$\dot{m} = f(LR\mu\Delta P)$
Large vertical opening with bidirectional flow	$\dot{m} = f(\rho HWH_rC_d\Delta P)$
General flow conduit (duct or pipe)	$\dot{m} = f(D_hALk\sum C_i v \Delta P)$
General flow inducer (fan or pump)	$\Delta P = \sum\limits_{i=0}^{3} a_i\left(\dfrac{\dot{m}}{\rho}\right)^i$
	$\dot{q}_{min} \leq \dfrac{\dot{m}}{\rho} \leq \dot{q}_{max}$
General flow corrector (damper or valve)	$\dot{m} = f(\rho_0 \Delta P_0 k_{vs} k_{v0} k_{vr} H/H_{100})$
	$H/H_{100} = f(daytimeS_1H_1S_uH_u)$
Flow corrector with polynomial local loss factor	$\dot{m} = f(\rho A \Delta P C)$
	$C = \sum\limits_{i=0}^{3} a_i\left(\dfrac{H}{H_{100}}\right)^i$
	$H/H_{100} = f(daytimeS_1H_1S_uH_u)$

component and supplementary data which depends on the flow component selected. Note that more than one connection may exist between two nodes. The concept of a connection having a positive side and a negative side is used to keep track of the direction of fluid flow. For most mass flow component types, unidirectional fluid flow will result (in either direction). However, some component types may represent bidirectional fluid movement—for example in the case of a doorway where, due to the action of small density variations over the height, bidirectional flow may exist.

4.2.1 The calculation process

Consider Figure 4.6 which shows two zones connected by some fluid flow component. It is assumed that each volume can be characterized by a single temperature and a single static pressure at some height relative to a common datum plane. The inlet and outlet of the connecting component are at different heights relative to each other and relative to the nodes representing the volumes. Analysis of the fluid flow through a component i is based on Bernoulli's Equation for one-dimensional steady flow of an incompressible Newtonian fluid including a loss term:

$$\Delta P_i = \left(p_1 + \frac{\rho v_1^2}{2}\right) - \left(p_2 + \frac{\rho v_2^2}{2}\right) + \rho g(z_1 - z_2)\ (\text{Pa}) \qquad (4.4)$$

where ΔP_i is the sum of all friction and dynamic losses (Pa); p_1, p_2, the entry and exit static pressures (Pa); v_1, v_2, the entry and exit velocities (m/s); ρ, the density of the fluid flowing through the component (kg/m^3); g, the acceleration of gravity (m/s^2); and z, the entry and exit elevation (m).

Bernoulli's Equation can be simplified by combining several related terms. Stack effects are represented by the $\rho g(z_1 - z_2)$ term in Equation (4.4). Dynamic pressures

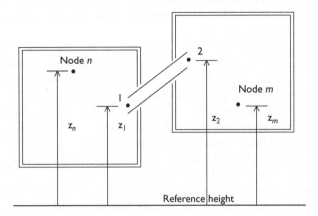

Figure 4.6 An example two zone connected system.

are the $\rho v^2/2$ terms, and total pressure is defined to be the sum of static pressure and dynamic pressure; that is, $P = p + (\rho v^2)/2$. If nodes n and m represent large volumes (e.g. a room), the dynamic pressures are effectively zero. If the nodes represent some point in a duct or pipe network, there will be a positive dynamic pressure. Equation (4.4) thus reduces to

$$\Delta P = P_n - P_m + PS_{nm} \text{ (Pa)} \tag{4.5}$$

where P_n, P_m are the total pressure at nodes n and m (Pa); and PS_{nm}, the pressure difference due to density and height differences across connection n through m (Pa).

Equations (4.4) and (4.5) define a sign convention for the direction of flow: positive from point 1 to point 2 (or n to m). The flow within each fluid flow component is described by a relation of the form $\dot{m} = f(\Delta P)$. The partial derivatives needed for the establishment of the Jacobian matrix (representing nodal pressure corrections in terms of all branch flow partial derivatives) are thus related by $\partial \dot{m}/\partial \Delta P_{nm} = -\partial \dot{m}/\partial \Delta P_{mn}$.

4.2.2 Flow calculation

As an example of flow calculation, consider the power law component types (A, B, or C). These flow components use one of the following relationships between flow and pressure difference across the component:

Type A: $\dot{m} = \rho a \Delta P^b \text{ (kg/s)}$ $\hspace{2cm}$ (4.6a)

Type B: $\dot{m} = a \Delta P^b \text{ (kg/s)}$ $\hspace{2cm}$ (4.6b)

Type C: $\dot{m} = a\sqrt{\rho} \Delta P^b \text{ (kg/s)}$ $\hspace{2cm}$ (4.6c)

where $\dot{m}$ is the fluid mass flow rate through the component (kg/s); a, the flow coefficient, expressed in m^3/s Pa^b (type A), kg/s Pa^b (type B), (kg m^3)$^{1/2}$ /s Pa^b (type C) ΔP, the total pressure loss across the component (Pa); and b, the flow exponent.

As can be seen, the difference between the three subtypes is only in the dimension of the flow coefficient a. Although in the literature all three forms can be found, the first one is the most commonly encountered.

The value of ρ depends on the type of fluid and on the direction of flow. If the flow is positive (i.e. when $\Delta P \geq 0$) then the temperature of the node on the positive side is used to evaluate the fluid density. Likewise, for a negative flow the temperature of the node on the negative side of the connection is used. Theoretically, the value of the flow exponent b should lie between 0.5 (for fully turbulent flow) and 1.0 (for laminar flow). The power law relationship should, however, be considered a correlation rather than a physical law. It can conveniently be used to characterize openings for building air infiltration calculations, because the majority of building fabric leakage description data is available in this form (Liddament 1986).

The power law relationship can also be used to describe flows through ducts and pipes. The primary advantage of the power law relationship for describing fluid flow components is the simple calculation of the partial derivative needed for the Newton–Raphson approach:

$$\frac{\partial \dot{m}}{\partial \Delta P} = \frac{b\dot{m}}{\Delta P} \text{ (kg/s/Pa)} \tag{4.7}$$

There is a problem with this equation however: the derivative becomes undefined when the pressure drop (and the flow) approach zero. This problem can be solved by switching to numerical approximation of the partial derivative in cases where the pressure drop is smaller than a certain threshold (say 10^{-20} Pa):

$$\frac{\partial \dot{m}}{\partial \Delta P} \approx \frac{\dot{m} - \dot{m}^*}{\Delta P - \Delta P^*} \text{ (kg/s/Pa)} \tag{4.8}$$

where * denotes the value in the previous iteration step.

4.2.3 Network solution

Each fluid flow component, i, thus relates the mass flow rate, $\dot{m}_i$, through the component to the pressure drop, ΔP_i, across it. Conservation of mass at each internal node is equivalent to the mathematical statement that the sum of the mass flows must equal zero at such a node. Because these flows are nonlinearly related to the connection pressure difference, solution requires the iterative processing of a set of simultaneous nonlinear equations subjected to a given set of boundary conditions. One technique is to assign an arbitrary pressure to each internal node to enable the calculation of each connection flow from the appropriate connection equation. The internal node mass flow residuals are then computed from

$$R_i = \sum_{k=1}^{K_{i,i}} \dot{m}_k \text{ (kg/s)} \tag{4.9}$$

where R_i is the node i mass flow residual for the current iteration (kg/s); $\dot{m}_k$, the mass flow rate along the kth connection to the node i (kg/s); and $K_{i,i}$, the total number of connections linked to node i.

The nodal pressures are then iteratively corrected and the mass balance at each internal node is reevaluated until some convergence criterion is met. This method—as implemented in ESP-r—is based on an approach suggested by Walton (1989a,b).

The solution method is based on a simultaneous whole network Newton–Raphson technique, which is applied to the set of simultaneous nonlinear equations. With this technique a new estimate of the nodal pressure vector, $\mathbf{P}^*$, is computed from d the current pressure field, $\mathbf{P}$, via

$$\mathbf{P}^* = \mathbf{P} - \mathbf{C} \tag{4.10}$$

where $\mathbf{C}$ is the pressure correction vector.

$\mathbf{C}$ is computed from the matrix product of the current residuals $\mathbf{R}$ and the inverse $\mathbf{J}^{-1}$ of a Jacobian matrix which represents the nodal pressure corrections in terms of all branch flow partial derivatives:

$$\mathbf{C} = \mathbf{R}\mathbf{J}^{-1} \tag{4.11}$$

where $\mathbf{J}$ is the square Jacobian matrix (N^*N for a network of N nodes)

The diagonal elements of $\mathbf{J}$ are given by

$$J_{n,n} = \sum_{k=1}^{K_{n,n}} \left(\frac{\partial \dot{m}}{\partial \Delta P}\right)_k \quad \text{(kg/s Pa)} \tag{4.12}$$

where $K_{n,n}$, is the total number of connections linked to node n; and ΔP_k, the pressure difference across the kth link.

The off-diagonal elements of $\mathbf{J}$ are given by

$$J_{n,m} = \sum_{k=1}^{K_{n,m}} -\left(\frac{\partial \dot{m}}{\partial \Delta P}\right)_k \quad \text{(kg/s Pa)} \tag{4.13}$$

where $K_{n,m}$ is the number of connections between node n and node m. This means that—for internal nodes—the summation of the terms comprising each row of the Jacobian matrix are identically zero.

Conservation of mass at each internal node provides the convergence criterion. That is, if $\sum \dot{m} = 0$ for all internal nodes for the current system pressure estimate, the exact solution has been found. In practice, iteration stops when all internal node mass flow residuals satisfy some user-defined criteria.

To be able to handle occasional instances of slow convergence due to oscillating pressure corrections on successive iterations, a method as suggested by Walton (1989a,b) was adopted. Oscillating behavior is indicated graphically in Figure 4.7 for the successive values of the pressure at a single node. In the case shown each successive pressure correction is a constant ratio of the previous correction, that is $C_i = -0.5\,C_i^*$ (* denotes the previous iteration step value). In a number of tests the observed oscillating corrections came close to such a pattern. By assuming a constant ratio, it is simple to extrapolate the corrections to an assumed solution:

$$P_i = P_i^* - \frac{C_i}{1 - r} \quad \text{(Pa)} \tag{4.14}$$

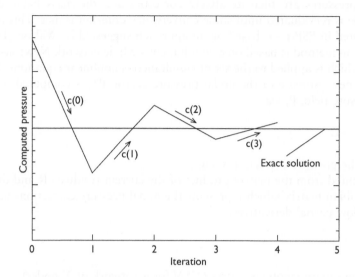

Figure 4.7 Example of successive computed values of the pressure and oscillating pressure corrections at a single node.

where r is the ratio of C_i for the current iteration to its value in the previous iteration. The factor $1/(1 - r)$ is called a relaxation factor. The extrapolated value of node pressure can be used in the next iteration. If it is used in the next iteration, then r is not evaluated for that node in the following iteration but only in the one thereafter. In this way, r is only evaluated with unrelaxed pressure correction values. This process is similar to a Steffensen iteration (Conte and de Boor 1972), which is used with a fixed-point iteration method for individual nonlinear equations. The iteration correction method presented here gives a variable and node-dependent relaxation factor. When the solution is close to convergence, Newton–Raphson iteration converges quadratically. By limiting the application of relaxation factor to cases where r is less than some value such as 0.5, it will not interfere with the rapid convergence.

However, there is some evidence that suggests that in a number of cases simple under relaxation would provide even better convergence acceleration than the Steffensen iteration (Walton 1990).

Some network simulation methods incorporate a feature to compute an initial pressure vector from which the iterations will start. For instance (Walton 1989a,b) uses linear pressure-flow relations for this. Reasons for refraining from this are as follows:

1   it is not possible to provide a linear pressure–flow relation for all envisaged flow component types;
2   after the initial start, the previous time step results probably provide better iteration starting values than those resulting from linear pressure–flow relations; and
3   this would impose an additional input burden upon the user.

According to Walton (1990) and Axley (1990), an initial pressure vector would also be necessary for low flow velocities so that (a) flows are realistically modeled in the laminar flow regimes, and (b) to avoid singular or nearly singular system Jacobians when employing Newton–Raphson solution strategies. Alternative solutions for these problems are (a) enable the problematic flow component types to handle laminar flow, and (b) to use a robust matrix solver.

4.3 Zonal modeling of coupled heat and airflow

In building energy prediction it is still common practice to separate the thermal analysis from the estimation of air infiltration and ventilation. This might be a reasonable assumption for many practical problems, where the airflow is predominantly pressure driven; that is wind pressure, or pressures imposed by the HVAC system. However, this simplification is not valid for cases where the airflow is buoyancy driven; that is, involving relatively strong couplings between heat and airflow. Passive cooling by increasing natural ventilation to reduce summertime overheating is a typical example.

Given the increased practical importance of such applications, there is a growing interest among building professionals and academics to establish prediction methods which are able to integrate air infiltration and ventilation estimation with building thermal simulation (Heidt and Nayak 1994).

Starting from the observation that it is not very effective to set up single equations describing both air and heat flow,[1] we see in practical applications two basic approaches for integrating or coupling a thermal model with a flow model:

1 the thermal model calculates temperatures based on assumed flows, after which the flow model recalculates the flows using the calculated temperatures, or
2 the flow model calculates flows based on assumed temperatures, after which the thermal model recalculates the temperatures using the calculated flows.

This means that either the temperatures (case 2) or the flows (case 1) may be different in both models, and steps need to be taken in order to ensure the thermodynamic integrity of the overall solution.

In the case where the thermal model and the flow model are actually separate programs which run in sequence, this procedure cannot be done on a per time step basis. This is the so-called sequential coupling as described by Kendrick (1993) and quantified with case study material by Heidt and Nayak (1994).

For applications involving buoyancy-driven airflow, the thermodynamic integrity of the sequential coupling should be seriously questioned. For those type of applications relative large errors in predicted temperatures and flows may be expected when using intermodel sequential coupling.

In the case where the thermal and flow model are integrated in the same software system (Figure 4.8), this procedure is possible for each time step and thermodynamic integrity can be guarded by

1 a decoupled approach ("ping-pong" approach) in which the thermal and flow model run in sequence (i.e. each model uses the results of the other model in the previous time step),[2] and
2 a coupled approach (or "onion" approach) in which the thermal and flow model iterate within one time step until satisfactory small error estimates are achieved.

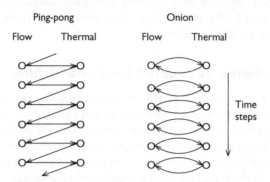

Figure 4.8 Schematic representations of decoupled noniterative ("ping-pong") and coupled iterative ("onion") approach.

Obviously, the final results in terms of evolution of the thermodynamic integrity will depend on how fast boundary values and other external variables to the models change over time. Therefore the length of the simulation time step is also an issue that needs to be considered.

In literature, several publications exist which relate to the modeling of coupled heat and air flow applications. Our own coupling approach has already been described earlier in detail (Clarke and Hensen 1991; Hensen 1991, 1999b), and is summarized in the next section.

Kafetzopoulos and Suen (1995) describe sequential coupling of the thermal program Apache with the airflow software Swifib. The results from both programs were transferred manually from one to the other, and this process was repeated until convergence to the desired accuracy was achieved. This procedure is very laborious, and so it was attempted for short simulation periods only.

Within the context of the IEA Energy Conservation in Buildings and Community Systems research, Dorer and Weber (1997) describes a coupling which has been established between the general purpose simulation package TRNSYS and the multi-zone airflow model COMIS.

Andre *et al.* (1998) report on usage of these coupled software packages. Initially, according to Andre (1998), the automatic coupling between the two software packages was not fully functional, so the results were transferred between the two programs in a way similar to the procedure followed by Kafetzopoulos and Suen (1995). However, as reported and demonstrated by Dorer and Weber (1999), the automatic coupling of the two software packages is now fully functional.

In all the above referenced works, the importance of accurate modeling of coupled heat and airflow is stressed, and in several cases demonstrated by case study material.

4.3.1   Implementation example

In order to generate quantitative results, it is necessary to become specific in terms of implementation of the solution methods. The work described in this section has been done with ESP-r, a general-purpose building performance simulation environment.

For modeling transient heat flow, this software uses a numerical approach for the simultaneous solution of finite volume energy conservation equations. For modeling airflow, the system features both a mass balance network approach and a CFD approach (Clarke 2001; Clarke *et al.* 1995). The former approach is used for the studies in this chapter.

In outline, the mass balance network approach involves the following: during each simulation time step, the mass transfer problem is constrained to the steady flow (possibly bidirectional) of an incompressible fluid (currently air and water are supported) along the connections which represent the building/plant mass flow paths network when subjected to certain boundary conditions regarding (wind) pressures, temperatures and/or flows. The problem therefore reduces to the calculation of airflow through these connections with the internal nodes of the network representing certain unknown pressures. A solution is achieved by an iterative mass balance technique (generalized from the technique described by Walton 1989a) in which the unknown nodal pressures are adjusted until the mass residual of each internal node satisfies some user-specified criterion.

Each node is assigned a node reference height and a temperature (corresponding to a boundary condition, building zone temperature or plant component temperature). These are then used for the calculation of buoyancy-driven flow or stack effect. Coupling of building heat flow and airflow models, in a mathematical/numerical sense, effectively means combining all matrix equations describing these processes.

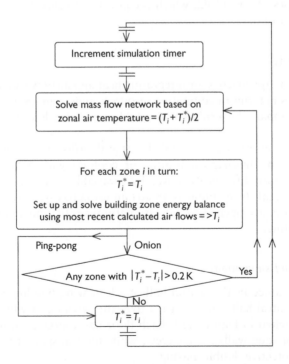

Figure 4.9 Schematic flow diagram showing the implementation of a coupled ("onion") and decoupled ("ping-pong") solution method for heat and airflow.

While, in principle, it is possible to combine all matrix equations into one overall "super-matrix", this is not done within this software, primarily because of the advantages that accrue from problem partitioning.

The most immediate advantage is the marked reduction in matrix dimensions and degree of sparsity—indeed the program never forms two-dimensional arrays for the above matrices, but instead holds matrix topologies and topographies as sets of vectors. A second advantage is that it is possible to easily remove partitions as a function of the problem in hand. For example, when the problem incorporates building-only considerations, plant-only considerations, plant + flow, and so on. A third advantage is that different partition solvers can be used which are optimized for the equation types in question—highly nonlinear, differential and so on.

It is recognized, however, that there often are dominating thermodynamic and/or hydraulic couplings between the different partitions. If a variable in one partition (say air temperature of a zone) depends on a variable of state solved within another partition (say the air flow rate through that zone), it is important to ensure that both values are matched in order to preserve the thermodynamic integrity of the system.

As schematically indicated in Figure 4.9, this can be achieved with a coupled ("onion") or decoupled ("ping-pong") solution approach. The flow diagram shows that in decoupled mode, within a time step, the airflows are calculated using the zonal air temperatures T_i of the previous time step; that is during the first pass through a time step, T_i equals T_i^* (history variable). In coupled mode, the first pass through a time step also uses the zonal air temperatures of the previous time step. However, each subsequent iteration uses $(T_i + T_i^*)/2$, which is equivalent to successive substitutions with a relaxation factor of 0.5.

4.3.2   Case study

Each of the various approaches for integrating heat and airflow calculations have specific consequences in terms of computing resources and accuracy. One way to demonstrate this is to compare the results for a typical case study (described in more detailed in Hensen 1995).

One of the most severe cases of coupled heat and airflow in our field involves a free running building (no mechanical heating or cooling) with airflow predominately driven by temperature differences caused by a variable load (e.g. solar load). A frequently occurring realistic example is an atrium using passive cooling, assuming that doors and windows are opened to increase natural ventilation so as to reduce summertime overheating.

4.3.2.1   Model and simulations

The current case concerns the central hall of a four-wing building located in central Germany. This central hall is in essence a five-story atrium, of which a cross-section and plan are sketched in Figure 4.10. Each floor has a large central void of $144\,\text{m}^2$. The floors and opaque walls are concrete, while the transparent walls and the roof consist of sun-protective double glazing.

In order to increase the infiltration, there are relatively big openings at ground and roof level. The eight building envelope openings ($2\,\text{m}^2$ each) are evenly distributed

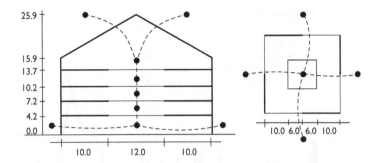

Figure 4.10 Cross-section and plan of atrium with airflow network (dimensions in m).

and connected as indicated in the flow network. For the present study, all openings are continuously open. Apart from solar gains, there are no other heat gains. There is no control (heating, cooling, window opening, etc.) imposed on the building.

The ambient conditions are taken from a weather test reference year for Wuerzburg, which is in the south-western part of Germany. The simulation period (28 August until 2 September) consists of a 6-day period with increasing outdoor air temperature to include a range of medium to maximum temperatures.

ESP-r features various modes of time step control. However, in order to avoid "interferences" which might make it difficult to interpret certain results in the current case, it was decided not to activate time step control. Instead of time step control, two time step lengths of respectively one hour and one-tenth of an hour were used during simulation.

4.3.2.2   Results and discussion

Figure 4.11 shows the simulation results for the vertical airflow through the atrium. In order to focus on the differences between the various methods, the right hand side of the figure shows two blown-up parts of the graphs. In the blown-ups, the different methods can clearly be distinguished. It can be seen that the ping-pong method with 1-h time steps is clearly an outlier relative to the other cases. For the 6-min time steps, the onion and ping-pong approaches give almost identical results.

In general, the flows tend to be higher during the night, and becomes less during the day. This effect is less pronounced during the first day, which has relatively low ambient air temperatures and levels of solar radiation.

The air temperatures on the ground floor show very little difference between the various approaches. This is probably due to the fact that the incoming air temperature (= ambient) is equal in all cases and because of the large thermal capacity of the ground floor.

Figure 4.12 shows the simulation results for the air temperatures on the top floor. Here the general graph and the blown-ups show larger differences between the various approaches. This is due to the succession of differences occurring at the lower floors and due to the fact that the top floor has a much higher solar gain (via the transparent roof) than the other floors.

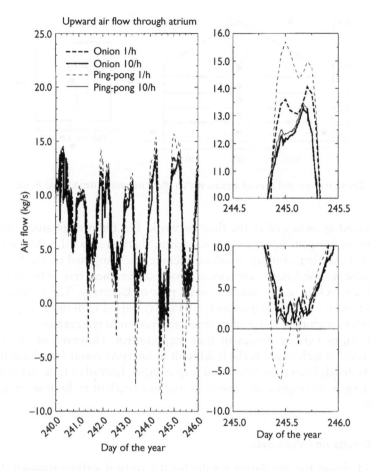

Figure 4.11 Simulation results for vertical airflow through atrium starting with day 241 and ending with day 246 of a reference year. Frames on the right show selected results in more detail. Flow rate in kg/s.

It is interesting to compare Figure 4.12 with Figure 4.11, because it shows that the flow increases with the difference between zonal and ambient temperatures and not with zonal temperature itself.

Obviously, the temperature difference depends on the amount of airflow, while the amount of airflow depends on temperature difference. As is clearly shown in the graphs, it takes an integrated approach to predict the net result.

Table 4.3 shows a statistical summary of the results. Included are the numbers of hours above certain temperature levels, since such parameters are used in certain countries to assess summer overheating. For the ground floor air temperatures there are relative big differences in hours >27°C between the once per hour and the 10 per hour time step cases. This is because the maximum air temperature for that zone is close to 27°C and so the number of hours above 27°C is very sensitive.

This case study focuses on the relative comparison of methodologies to model coupled heat and airflow in a building. Although no mathematical proof is presented,

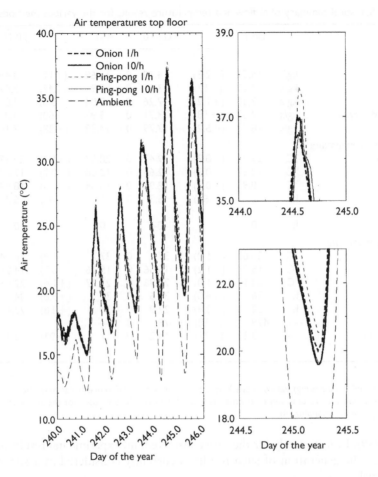

Figure 4.12 Simulation results for top floor air temperatures starting with day 241 and ending with day 246 of a reference year. Frames on the right show selected results in more detail (temperatures in °C).

it could be argued that in the current situation the results for the coupled solution method with small time steps are the most accurate. This is why for each result the percentage difference is shown relative to the results for the coupled solution with 10 time steps per hour.

Since the main interest here are the relative differences, no attempt has been made to compare the case study results by intermodel comparison, for example with a CFD approach, or to validate the outcome in an absolute sense by comparing with experimental results.

A comparison with CFD results would not constitute a feasible option because modeling of coupled building energy and CFD is still very much in its infancy (Beausoleil-Morrison 2000; Zhai *et al.* 2002; Djunaedy *et al.* 2003).

Each of the decoupled building energy and airflow prediction methods have been subjected to extensive and rigorous experimental validation exercises in the past

Table 4.3 Statistical summary of airflow and temperature results for the various methods

		On-1		On-10		PP-1		PP-10	
Vertical flow									
Maximum	kg/s	14.51	(+2.3)	14.19	0	15.69	(+11)	13.49	(−4.9)
Minimum	kg/s	−4.21	(+17)	−3.6	0	−8.9	(+247)	−3.67	(+1.9)
Mean	kg/s	7.35	(+1.2)	7.26	0	7.04	(−3.0)	7.05	(−2.9)
Standard deviation	kg/s	4.37	(+18)	3.71	0	5.93	(+60)	3.87	(+4.3)
Range	kg/s	18.72	(+5.2)	17.79	0	24.58	(+38)	17.16	(−3.5)
Ground floor temperature									
Maximum	°C	29.21	(−0.7)	29.42	0	28.87	(−1.7)	29.37	(−0.2)
Minimum	°C	12.67	(+0.3)	12.63	0	12.66	(+0.2)	12.63	(+0.0)
Mean	°C	18.95	(+0.1)	18.93	0	18.64	(−1.5)	18.84	(−0.5)
> 27°C	h	2	(−62)	5.3	0	1	(−81)	6.3	
(+19)									
> 30°C	h	0		0		0		0	
Top floor temperature									
Maximum	°C	36.63	(−1.0)	37	0	37.7	(+1.9)	36.94	(−0.2)
Minimum	°C	15.24	(+1.2)	15.06	0	15.16	(+0.7)	14.91	(−1.0)
Mean	°C	23.19	(+1.0)	22.96	0	23.27	(+1.4)	22.83	(−0.6)
>27°C	h	36	(+4.0)	34.6	0	38	(+9.8)	34.3	(−0.9)
>30°C	h	22	(−3.9)	22.9	0	24	(+4.8)	23.4	(+2.2)
Iterations	—	429	(−58)	1,028	0	—		—	
Relative user CPU	—	3.3	(−77)	14.2	0	1	(−93)	8.3	(−41)

Notes
(On = onion, PP = ping-pong). Values in brackets are the percentage differences relative to the On-10 case, that is, coupled solution with 10 time steps per hour. User CPU time, is CPU time used for the actual calculations, that is, excluding time for swapping, etc.

(CEC 1989). Unfortunately for the case considered, no experimental results are readily available. The generation of such results is currently considered as a suggestion for future work.

The largest discrepancies between the various coupling methods are found for case PP-1, that is decoupled solution with relatively large time steps. The results for the coupled solution cases and for the decoupled solution with small time steps are relatively close.

Table 4.3 also shows the number of iterations needed for each case with the coupled solution approach. The amount of code involved in the iteration is only a fraction of the code that needs to be processed for a complete time step.

In terms of computer resources used, it is more relevant to compare the user CPU time as shown at the bottom of Table 4.3. The results are shown relative to the PP-1 case, which was the fastest method. It is clear that the other cases use much more computer resources; especially the coupled solution method with small time steps.

4.3.2.3   Case study conclusions

The case study presented here involves a case of strongly coupled heat and air flow in buildings. Two different methods, that is coupled and decoupled solutions, for linking heat and airflow models have been considered using two different time step lengths.

It was found that the differences are much larger in terms of airflow than in terms of air temperatures. The temperature differences between the various methods increases with the number of stacked zones.

The main conclusion from the case study is that the coupled solution method will be able to generate accurate results, even with simulation time steps of 1 h. Reducing the time step will increase the computing resources used considerably, with a relatively small improvement of the accuracy.

For equal length of time steps a coupled solution method will use more computer resources than a decoupled solution.

For the decoupled method, it is necessary to reduce the time step to ensure the accuracy. For the current case study, the decoupled solution method using a simulation time step of 360 s was less accurate than the coupled solution method with a time step of 1 h. However, the computer resources used were more than doubled.

Based on the current case study, it may be concluded that the coupled solution gives the best overall results in terms of both accuracy and computer resources used. Although the results presented here are for an imaginary (but realistic) building, the observed trends may be expected to be more generally valid.

4.4   Quality assurance

Due to lack of available resources it usually has to be assumed in a practical design study context that the models and the simulation environment, which is being used, has been verified (i.e. the physics are represented accurately by the mathematical and numerical models) and validated (i.e. the numerical models are implemented correctly). Nevertheless, it is critically important to be aware of the limitations of each modeling approach.

For example, when using the network approach it should be realized that most of the pressure–flow relationships are based on experiments involving turbulent flow. Von Grabe *et al.* (2001) demonstrate the sensitivity of temperature rise predictions in a double-skin façade, and the difficulty of modeling the flow resistance of the various components. There are many factors involved but assuming the same flow conditions for natural ventilation as those used for mechanical ventilation causes the main problem, that is using local loss factors ζ and friction factors from mechanical engineering tables. These values have been developed in the past for velocities and velocity profiles as they occur in pipes or ducts: symmetric and having the highest velocities at the center. With natural ventilation however, buoyancy is the driving force. This force is greater near the heat sources, thus near the surface and the shading device, which will lead to nonsymmetric profiles. This is worsened because of the different magnitudes of the heat sources on either side of the cavity.

One way forward would be to use CFD in separate studies to predict appropriate local loss factors ζ and friction factors for use in network methods. Strigner and Janak (2001) describe an example of such a CFD approach by predicting the aerodynamic performance of a particular double-skin façade component, an inlet grill. However, as indicated earlier, CFD is still very much being developed. At the same time it seems to be very appealing to engineers and clients; the CFD = colors for directors effect? Therefore it is essential that quality assurance procedures such as by Chen and Srebric (2001) will be developed.

In some occasions, such as buoyancy-driven flow in complex networks comprising both very small and very large airflow openings, airflow oscillations may occur. This may be because buoyancy and other forces are almost in balance, in reality as well as in the model, thus the flow is close to unstable which can result in oscillations. One of the reasons why it may not happen in reality but does happen in the simulations is that in the energy balance and flow network approach "only" energy and mass conservation are taken into account. The momentum of the flow is not considered.

If such a situation occurs (i.e. flow close to unstable in reality) more (onion) iterations or smaller time steps will not help to avoid oscillations. Smaller time steps will however reduce the "amplitude" of the oscillations, and—if required—will avoid unstable oscillations. It is, by the way, very easy to construct an unrealistic airflow network without being aware of it!

As discussed elsewhere in more detail (Hensen 1999a) another limitation is related to assumed ambient conditions. This concerns the difference between the "micro climate" near a building and the weather data, which is usually representative of a location more or less distant from the building. These differences are most pronounced in terms of temperature, wind speed and direction, the main driving potential variables for the heat and mass transfer processes in buildings!

These temperature differences are very noticeable when walking about in the summer in an urban area. Yet it seems that hardly any research has been reported or done in this area. There are some rough models to predict the wind speed reduction between the local wind speed and the wind speed at the meteorological measurement site. This so-called wind speed reduction factor accounts for any difference between measurement height and building height and for the intervening terrain roughness. It assumes a vertical wind speed profile, and usually a stable atmospheric boundary layer.

It should be noted however that most of these wind profiles are actually only valid for heights over $20z_0 + d$ (z_0 is the terrain-dependent roughness length (m), and d is the terrain-dependent displacement length (m)) and lower than 60–100 m; that is, for a building height of 10 m in a rural area, the profiles are only valid for heights above 17 m, in an urban area above 28 m and in a city area above 50 m. The layer below $20z_0 + d$ is often referred to as the urban canopy. Here the wind speed and direction is strongly influenced by individual obstacles, and can only be predicted through wind tunnel experiments or simulation with a CFD model. If these are not available, it is advised to be very cautious, and to use—depending on the problem at hand—a high or low estimate of the wind speed reduction factor. For example, in case of an "energy consumption and infiltration problem" it is safer to use a high estimate of the wind speed reduction factor (e.g. wind speed evaluated at a height of $20z_0 + d$). In case of an "air quality" or "overheating and ventilation" problem it is probably safer to use a low estimate (e.g. wind speed evaluated at the actual building height, or assuming that there is no wind at all).

Calibration is a very difficult issue in practice. For existing buildings there are usually no experimental results readily available. In a design context there is not even a building yet. In practice, the only way to calibrate the model is to try to gain confidence by carefully analyzing the predictions and to compare these to expectations or "intuition" based on previous work. Unexpected results are usually the result of modeling errors. In rare—but interesting—cases unexpected interactions take place and—after analyzing these—the simulations may have helped to improve the understanding of the problem. In any event, calibration should not be taken lightly and sufficient resources should be reserved for this activity.

4.5   Performance assessment methodology

Simulation quality can only be assured through an appropriate performance assessment methodology. This should always include selection of the correct model resolution/complexity level and calibration as indicated earlier. Obviously simulations should be performed with relevant inside and ambient boundary conditions during a suitable length of time. The results should be thoroughly analyzed and reported. Next the model should be changed to reflect another design option, and the procedure of simulation, results analysis and reporting should be repeated until the predictions are satisfactory. It is very important to convey to clients that simulation is much better for performance-based relative rank-ordering of design options, than for predicting the future performance of a final design in absolute terms. It is "interesting" that this is more likely to be "forgotten" in higher resolution modeling exercises.

A good performance assessment methodology should also take into account the limitations of the approach, for instance by a min–max approach or by sensitivity analysis.

Too low resolution and/or too simplified approaches might not reliably solve a particular problem. On the other hand, approaches with too high resolution or too much complexity might lead to inaccuracy as well (although this statement cannot be substantiated yet). Obviously, too high resolution or too complex approaches will require excessive amount of resources in terms of computing capacity, manpower and time. How to select the appropriate approach to solve the problem at hand remains the challenge.

Slater and Cartmell (2003) developed what they called "early assessment design strategies" (Figure 4.13). From early design brief, the required complexity of the modeling can be assessed. Starting from the design standard, a building design can be assessed whether it falls safely within the Building Regulations criteria, or in the borderline area where compliance might fail, or in a new innovative design altogether. Based on this initial assessment, and with the proposed HVAC strategy, several decision points in Figure 4.13 will help the engineer to decide which level of complexity should be used for simulation. There is a need to go further than what Slater and Cartmell (2003) propose because of the following reasons:

1   Coupled approaches (between energy simulation and CFD) will soon be a viable option that is not addressed in Figure 4.13.
2   As Hensen et al. (1996) point out, we need to use different levels of complexity and resolution at different stages of building design.

Figure 4.14 shows a prototype performance-based airflow modeling selection strategy (AMS). This prototype was initially developed as part of a research on coupled simulation between CFD and building energy simulation (BES) (Djunaedy et al. 2003). The fact that coupled CFD and BES is computationally expensive requires that the simulation is justified by a sound rationale. AMS was proposed to identify the need for coupled simulation. However, AMS can be applied more generally to assess the need for a certain level of complexity and resolution for simulation.

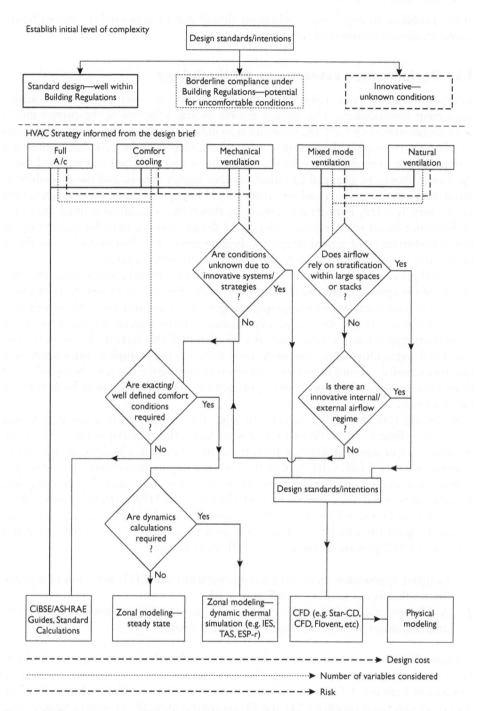

Figure 4.13 Early assessment design strategies (adapted from Slater and Cartmell 2003).

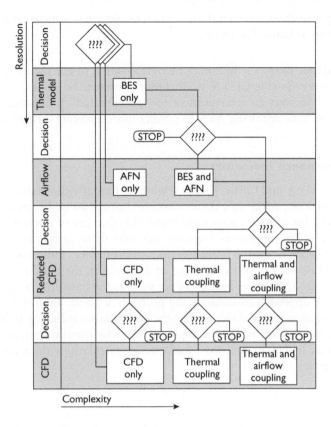

Figure 4.14 Prototype performance-based airflow modeling selection strategy.

The main ideas behind AMS are as follows:

1 A simulation should be consistent with its objective, that is the designer should not be tool-led.
2 There should be a problem-led rationale to progress from one level of resolution and complexity to the next.
3 Simulation should be made at the lowest possible resolution and complexity level, so that later there will be less design options to be simulated at higher resolution level.

On the vertical axis of Figure 4.14 there are layers of different resolution of building simulation. The four layers of increasing level of resolution are building energy simulation, airflow network simulation, and CFD simulation. One or more decision layers separate each of the resolution layers. The horizontal axis shows the different levels of complexity in building airflow simulation.

The first step is to select the minimum resolution based on the design question at hand. For example

• If energy consumption is needed, then BES would be sufficient.
• If temperature gradient is needed, then at least an AFN (Air Flow Network) is required.
• If local mean temperature of air is in question, then CFD is necessary.

A second step is to check whether the minimum resolution is sufficiently accurate for the design question at hand. For example

- Load analysis based on BES may be oversensitive to convection coefficient (hc) values, thus requiring CFD to predict more accurate hc values.
- Load analysis may be oversensitive to "guestimated" infiltration or interzonal ventilation, thus requiring AFN to predict more accurate airflow rates.

4.5.1   Performance indicators

Different from Slater and Cartmell (2003) who use the early design brief as the base for the decision-making, AMS uses performance indicators to make decisions. Table 4.4 shows a typical list of performance indicators (PI) that are of interest for an environmental engineer. The indicators basically fall into three categories, that is, energy-related, load-related, and comfort-related performance indicators. Each of the categories will be used for different kind of decisions in the building design process.

With regard to AMS, these indicators are used as the basis for the decision to select the appropriate approach to simulate the problem at hand. Table 4.4 also shows the minimum resolution required to calculate the performance indicator. It should be noted that this is case dependent. For example in case of naturally ventilated double-skin façade, such as in Figure 4.3, load and energy calculations do require an airflow network approach.

Table 4.4 Example of performance indicators and (case dependent) minimum approach in terms of modeling resolution level

Performance Indicators	Approach
Energy related	
Heating energy demand	BES
Cooling energy demand	BES
Fan electricity	BES
Gas consumption	BES
Primary energy	BES
Load related	
Maximum heating load	BES
Maximum cooling load	BES
Comfort related	
PPD	BES
Maximum temperature in the zone	BES
Minimum temperature in the zone	BES
Over heating period	BES
Local discomfort, temperature gradient	AFN
Local discomfort, turbulence intensity	CFD
Contaminant distribution	AFN
Ventilation efficiency	AFN
Local mean age of air	CFD

4.5.2 Sensitivity analysis

Sensitivity analysis is the systemic investigation of the reaction of the simulation response to either extreme values of the model's quantitative factors or to drastic changes in the model's qualitative factors (Kleijnen 1997). This analysis has been used in many fields of engineering as a what-if analysis, and one example of the use of this method in building simulation is given by Lomas and Eppel (1992).

The main use of sensitivity analysis is to investigate the impact of a certain change in one (or more) input to the output. Depending on the particular problem, the end result is usually to identify which input has the most important impact on the output.

It has long been recognized as an essential part in model verification and/or validation. Recently several authors (Fuhrbringer and Roulet 1999; de Wit 2001; MacDonald 2002) suggested that sensitivity analysis should also be used in performing simulations.

For AMS, the sensitivity analysis will be used for yet another purpose, as the objective is not to identify which input is important, but rather to identify the effect of changes in a particular input on a number of outputs.

From previous studies, for example, Hensen (1991), Negrao (1995), and Beausolleil-Morrison (2000), we know that there are two main inputs that should be tested for sensitivity analysis for the decision whether to progress to higher resolution level:

- Airflow parameters assumption, especially the infiltration rate, for the decision whether to use AFN-coupled simulation. (Further sensitivity analysis on airflow parameters would be denoted as SA_{af}.)
- Convection coefficient, for the decision to use CFD. (Further sensitivity analysis on airflow parameters would be denoted as SA_{hc}.)

Figure 4.15 shows two scenarios on how to use AMS. Each of the performance indicators would have a "target value" that can be found from building codes, standards, or guidelines, or even from "good-practices" experience. The target value can be a maximum value, minimum value, or a range of acceptable values. The result of the sensitivity analysis would be presented as a bar chart with three output conditions of the performance indicator, corresponding to the minimum value, maximum value and base value of the input parameter.

In Figure 4.15(a), the output value is higher than the maximum target value, based on the result of BES-only simulation, and so the SA_{af} result indicates that the AFN-coupling is necessary. However, on the AFN-level, the SA_{hc} result indicates that all predicted values are below the maximum target value, thus no subsequent CFD calculation is required.

In Figure 4.15(b), the output value could be less than the minimum target value, based on the result of BES-only simulation, and so the SA_{af} result indicates that the AFN-coupling is necessary. On the AFN-level, the SA_{hc} result indicates that there is a possibility that the output value is below the minimum target value, thus CFD calculation will be required.

In conclusion, AMS suggests a rationale for selecting appropriate energy and airflow modeling levels for practical design simulations, which

- reduces the number of design alternatives to be considered at higher levels of resolution;
- focuses in terms of simulation periods at higher levels of resolution;

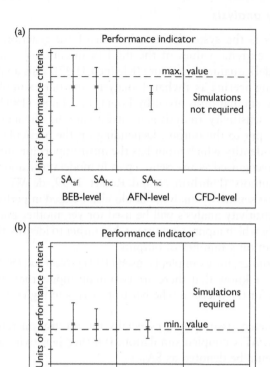

Figure 4.15 Different scenarios resulting from sensitivity analysis in AMS.

- indicates whether (de)coupled BES/CFD simulation will be needed; and
- constitutes a prerequisite for working out the mechanics of (external) coupling of BES/AFN/CFD.

As indicated here, both the coupling and the methodology work are still in the early phases.

4.6   Conclusion

Although much progress has been made there remain many problematic issues in building airflow simulation. Each modeling approach, from the semi-empirical and simplified methods to CFD, suffers from shortcomings that do not exist—or are much less—in other methods.

Also in terms of performance prediction potential, there is no single best method. Each method has its own (dis)advantages. Which method to use depends on the type of analysis that is needed at a particular stage in the design process. A simulation quality assurance procedure is very important. Apart from the essential need for domain knowledge, parts of such procedure might be semi-automated; see for example, Djunaedy *et al.* (2002). This is another interesting direction for future work.

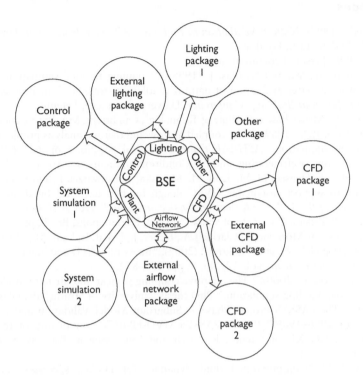

Figure 4.16 A future integrated building simulation environment based on an advanced multizone building performance simulation system which is run-time coupled to external software packages.

Although most of the basic physical models for airflow in and around buildings are accessible by today's computational techniques, there is still a lot of effort necessary until they can be widely used for problems in engineering practice. Moreover, the desired integration of algorithms with efficient data structures and adequate modeling techniques supporting the cooperation of partners in the design process still at a very premature stage.

As schematically shown in Figure 4.16 and elaborated in Djunaedy *et al.* (2003), one way forward could be via run-time coupling of distributed applications (as opposed to integration by merging code) which would enable multi-level modeling (the modeling and simulation laboratory metaphor), and will allow task-shared development of building performance simulation tools and techniques.

Notes

1 Other opinions exist (see e.g. Axley and Grot 1989), single equations describing both air and heat flow are sometimes referred to as "full integration" (Kendrick 1993).
2 In Figure 4.8 the airflow calculations use air temperatures calculated in the previous time step. Obviously the other way around is also possible.

References

Allard, F. (ed.) (1998). *Natural Ventilation in Buildings: A Design Handbook*. James & James Science Publishers Ltd., London.

Andre, P. (1998). Personal communication.

Andre, P., Kummert, M., and Nicolas, J. (1998). "Coupling thermal simulation and airflow calculation for a better evaluation of natural ventilation strategies." In *Proceedings of the System Simulation in Buildings Conference*. University of Liege, Belgium.

Axley, J.W. (1990). Massachusetts Institute of Technology, Cambridge, MA. Private communication.

Axley, J. and Grot, R.A. (1989). "The coupled airflow and thermal analysis problem in building airflow system simulation." *ASHRAE Transactions*, Vol. 95, No. 2, pp. 621–628, American Society of Heating, Refrigerating, and Air-Conditioning Engineers, Atlanta, GA.

Bartak, M., Drkal, F., Hensen, J.L.M., Lain, M., Matuska, T., Schwarzer, J., and Sourek, B. (2001). "Simulation to support sustainable HVAC design for two historical buildings in Prague." In *Proceedings of the 18th Conference on Passive and Low Energy Architecture*. PLEA 2001, November 7–9, Federal University of Santa Catarina, Florianopolis, Brazil, pp. 903–908.

Bartak, M., Beausoleil-Morrison, I., Clarke, J.A., Denev, J., Drkal, F., Lain, M., Macdonald, I.A., Melikov, A., Popiolek, Z., and Stankov, P. (2002). "Integrating CFD and building simulation." *Building and Environment*, Vol. 37, No. 8, pp. 865–872.

Beausoleil-Morrison, I. (2000). "The adaptive coupling of heat and air flow modeling within dynamic whole-building simulation." PhD thesis. University of Strathclyde, Glasgow, UK.

CEC (1989). "The PASSYS Project Phase 1. Subgroup Model Validation and Development Final Report 1986–1989." 033-89-PASSYS-MVD-FP-017, Commission of the European Communities, DG XII of Science, Research and Development, Brussels. S.O/ stergaard Jensen (ed.).

Chen, Q. (1997). "Computational fluid dynamics for HVAC: successes and failures." *ASHRAE Transactions*, Vol. 103, No. 1, pp. 178–187, American Society of Heating, Refrigerating, and Air-Conditioning Engineers, Atlanta, GA.

Chen, Q. and Srebic, J. (2000). "Application of CFD tools for indoor and outdoor environment design." *International Journal on Architectural Science*, Vol. 1, No. 1, pp. 14–29.

Chen, Q. and Srebic, J. (2001). "How to verify, validate and report indoor environmental modeling CFD analysis." *ASHRAE report RP-1133*, American Society of Heating, Refrigerating, and Air-Conditioning Engineers, Atlanta, GA.

Clarke, J.A. (2001). *Energy Simulation in Building Design*, 2nd edition. Butterworth Heinemann, Oxford.

Clarke, J.A. and Hensen, J.L.M. (1991). "An approach to the simulation of coupled heat and mass flow in buildings." In *Proceedings of the 11th AIVC Conference on Ventilation System Performance*. Belgirate (I) 1990, Vol. 2, pp. 339–354, IEA Air Infiltration and Ventilation Centre, Coventry (UK).

Clarke, J.A., Hensen, J.L.M., and Negrao, C.O.R. (1995). "Predicting indoor airflow by combining network, CFD, and thermal simulation." In *Proceedings of the 16th AIVC Conference "Implementing the Results of Ventilation Research."* Palm Springs, September, pp. 145–154, IEA Air Infiltration and Ventilation Centre, Coventry, UK.

COMIS Multizone airflow model: conjunction of multizone infiltration specialists, online available <http://epb1.lbl.gov/EPB/comis/> (accessed 23 September 2003).

CONTAMW Multizone airflow and contaminant transport analysis software, online available <http://www.bfrl.nist.gov/IAQanalysis/CONTAMW> (accessed 23 September 2003).

Conte, S.D. and de Boor, C. (1972). *Elementary Numerical Analysis: An Algorithmic Approach*. Mc-Graw-Hill, New York.

Crouse, B., Krafczyk, M., Kuhner, S., Rank E., and van Treeck, C. (2002). "Indoor airflow analysis based on lattice Boltzmann methods." *Energy and Buildings*, Vol. 34, No. 9, pp. 941–949.

Djunaedy, E., Hensen, J.L.M., and Loomans, M.G.L.C. (2002). "A strategy for integration of CFD in building design." Roomvent 2002, *Proceedings of 8th International Conference on Air Distribution in Rooms*, pp. 363–396, Copenhagen.

Djunaedy, E., Hensen, J.L.M., and Loomans, M.G.L.C. (2003). "Towards external coupling of building energy and airflow modeling programs." *ASHRAE Transactions*, Vol. 102, No. 2, American Society of Heating, Refrigerating, and Air-Conditioning Engineers, Atlanta, GA.

Dorer, V. and Weber, A. (1997). "Multizone air flow model COMVEN-TRNSYS, TRNSYS Type 157." *IEA-ECB Annex 23 report*, EMPA, Swiss Federal Laboratories for Materials Testing and Research.

Dorer, V. and Weber, A. (1999). "Air, Contaminant and Heat Transport Models: Integration and Application." *Energy and Buildings*, Vol. 30, No. 1, pp. 97–104.

ESP-r Integrated building performance simulation tool, online available <http://www.esru.strath.ac.uk/Programs/ESP-r.htm> (accessed 23 September 2003).

Etheridge, D.W. and Sandberg, M. (1996). *Building Ventilation: Theory and Measurement*. John Wiley & Sons Ltd., Chichester, England.

Fuhrbringer, J.-M. and Roulet, C.-A. (1999). "Confidence of simulation results: put a sensitivity analysis module in your MODEL: the IEA-ECBCS Annex 23 experience of model evaluation." *Energy and Buildings*, Vol. 30, No. 1, pp. 61–71.

Grabe, J. von, Lorenz, R., and Croxford, B. (2001). "Ventilation of double facades." In *Proceedings of the Building Simulation 2001*. Rio de Janeiro, pp. 229–236, International Building Performance Simulation Association.

Haghighat, F., Wang, J.C.Y., Jiang, Z., and Allard, F. (1992). "Air movement in buildings using computational fluid dynamics." *Transactions of the ASME Journal of Solar Energy Engineering*, Vol. 114, pp. 84–92.

Heidt, F.D. and Nayak, J.K. (1994). "Estimation of air infiltration and building thermal performance." *Air Infiltration Review*, Vol. 15, No. 3, pp. 12–16.

Hensen, J.L.M. (1990). "ESPmfs, a building & plant mass flow network solver." Collaborative FAGO/ESRU report 90.14.K, Eindhoven University of Technology.

Hensen, J.L.M. (1991). "On the thermal interaction of building structure and heating and ventilating system." Doctoral dissertation. Technische Universiteit Eindhoven, Netherlands.

Hensen, J.L.M. (1994). "Predictive modelling of air humidity in Glasgow's Peoples Palace." ESRU Report R94/20, University of Strathclyde, Energy Systems Research Unit.

Hensen, J.L.M. (1995). "Modelling coupled heat and airflow: ping-pong vs onions." In *Proceedings of the 16th AIVC Conference "Implementing the Results of Ventilation Research."* Palm Springs, September, pp. 253–262, IEA Air Infiltration and Ventilation Centre, Coventry (UK).

Hensen, J.L.M. (1999a). "Simulation of building energy and indoor environmental quality—some weather data issues." In *Proceedings of the International Workshop on Climate Data and Their Applications in Engineering*. 4–6 October, Czech Hydrometeorological Institute in Prague.

Hensen, J.L.M. (1999b). "A comparison of coupled and de-coupled solutions for temperature and air-flow in a building." *ASHRAE Transactions*, Vol. 105, No. 2, pp. 962–969, American Society of Heating, Refrigerating, and Air-Conditioning Engineers.

Hensen, J.L.M., Hamelinck, M.J.H., and Loomans, M.G.L.C. (1996). "Modelling approaches for displacement ventilation in offices." In *Proceedings of the 5th International Conference Roomvent '96*. Yokohama, July 1996, University of Tokyo.

Hensen, J.L.M., Bartak, M., and Drkal, F. (2002). "Modeling and simulation of a double-skin façade system." *ASHRAE Transactions*, Vol. 108, No. 2, pp. 1251–1259, American Society of Heating, Refrigerating, and Air-Conditioning Engineers, Atlanta, GA.

Jiang, Y. and Chen, Q. (2001). "Study of natural ventilation I buildings by large eddy simulation." *Journal of Wind Engineering and Industrial Aerodynamics*, Vol. 89, pp. 1155–1178.

Kafetzopoulos, M.G. and Suen, K.O. (1995). "Coupling of thermal and airflow calculation programs offering simultaneous thermal and airflow analysis." *Building Services Engineering Research and Technology*, Vol. 16, pp. 33–36.

Kendrick, J. (1993). "An overview of combined modelling of heat transport and air movement." *Technical Note AIVC 30*, Air Infiltration and Ventilation Centre, Coventry UK.

Kleijnen, J.P.C. (1997). "Sensitivity analysis and related analyses: a review of some statistical techniques." *Journal of Statistical Computation and Simulation*, Vol. 57, pp. 111–142.

Liddament, M. (1986). "Air infiltration calculation techniques—an applications guide." IEA Air Infiltration and Ventilation Centre, Coventry (UK).

Lomas, K.J. and Eppel, H. (1992). "Sensitivity analysis techniques for building thermal simulation programs." *Energy and Buildings*, Vol. 19, pp. 21–44.

MacDonald, I.A. (2002). "Quantifying the effects of uncertainty in building simulation." PhD thesis. University of Strathclyde, Glasgow, Scotland.

Martin, P. (1999). "CFD in the real world." *ASHRAE Journal*, Vol. 41, No. 1, pp. 20–25.

Negrao, C.O.R. (1995). "Conflation of computational fluid dynamics and building thermal simulation." PhD thesis. University of Strathclyde, Glasgow.

Orme, M. (1999). "Applicable models for air infiltration and ventilation calculations." IEA Air Infiltration and Ventilation Centre, *Technical Report 51*, Birmingham, UK.

Slater, S. and Cartmell, B. (2003). "Hard working software." *Building Services Journal*, Vol. 25, No. 2, February, pp. 37–40.

Strigner, R. and Janak, M. (2001). "Computer simulation of ventilated double-skin façade of the metropolitan library in Brno." In *Proceedings of the 1st International Conference on Renewable Energy in Buildings*. Sustainable Buildings and Solar Energy 2001, Brno, 15–16 November, Brno University of Technology Czech Academy of Sciences in Prague, pp. 134–137.

TRNSYS Transient systems simulation program, online available <http://sel.me.wisc.edu/trnsys/> (accessed 23 September 2003).

Walton, G.N. (1989a). "Airflow network models for element-based building airflow modeling." *ASHRAE Transactions*, Vol. 95, No. 2, pp. 613–620, American Society of Heating, Refrigerating, and Air Conditioning Engineers, Atlanta, GA.

Walton, G.N. (1989b.) "AIRNET—a computer program for building airflow network modeling." *NISTIR 89-4072*, National Institute of Standards and Technology, Gaithersburg, MD.

Walton, G.N. (1990). National Institute of Standards and Technology, Gaithersburg, MD. Private Communication.

de Wit, M.S. (2001). "Uncertainty in predictions of thermal comfort in buildings." PhD thesis. Technische Universiteit Delft, Netherlands.

Zhai, Z., Chen, Q., Haves, P., and Klems, J.H. (2002). "On approaches to couple energy simulation and CFD programs." *Building and Environment*, Vol. 37, No. 8, pp. 857–864.

Chapter 5

The use of Computational Fluid Dynamics tools for indoor environmental design

Qingyan (Yan) Chen and Zhiqiang (John) Zhai

5.1 Introduction

Since human beings spend more than 90% of their time indoors in developed countries, design of indoor environment is crucial to the comfort and welfare of the building occupants. However, this is not an easy task. Woods (1989) reported that about 800,000 to 1,200,000 commercial buildings in the United States containing 30–70 million workers have had problems related to the indoor environment. If the problems can be fixed through technologies, Fisk (2000) estimated that for the United States, the potential annual savings and productivity could be $15–$40 billion from reduced sick building syndrome symptoms, and $20–$200 billion from direct improvements in worker performance that are unrelated to health.

In addition, building safety is a major concern of building occupants. Smoke and fire has claimed hundreds of lives every year in the United States. After the anthrax scare following the September 11, 2001 attacks in the United States, how to protect buildings from terrorist attacks by releasing chemical/biological warfare agents becomes another major issue of building safety concerns.

In the past few years, Computational Fluid Dynamics (CFD) has gained popularity as an efficient and useful tool in the design and study of indoor environment and building safety, after having been developed for over a quarter of a century. The applications of CFD in indoor environment and building safety are very wide, such as some of the recent examples for natural ventilation design (Carriho-da-Graca *et al.* 2002), prediction of smoke and fire in buildings (Lo *et al.* 2002; Yeoh *et al.* 2003), particulate dispersion in indoor environment (Quinn *et al.* 2001), building element design (Manz 2003), and even for space indoor environment analysis (Eckhardt and Zori 2002). Some other applications are more complicated and may deal with solid materials, and may integrate other building simulation models. Recent examples are the study of building material emissions for indoor air quality assessment (Topp *et al.* 2001; Huang and Haghighat 2002; Murakami *et al.* 2003) and for more accurate building energy and thermal comfort simulations (Bartak *et al.* 2002; Beausoleil-Morrison 2002; Zhai and Chen 2003). Often, the outdoor environment has a significant impact on the indoor environment, such as in buildings with natural ventilation. To solve problems related to natural ventilation requires the study of both the indoor and outdoor environment together, such as simulations of outdoor airflow and pollutant dispersion (Sahm *et al.* 2002; Swaddiwudhipong and Khan 2002) and combined indoor and airflow studies (Jiang and Chen 2002). CFD is no longer a patent for users with PhD degrees. Tsou (2001) has developed online CFD as a teaching tool for building performance studies, including issues such as structural stability, acoustic quality, natural lighting, thermal comfort, and ventilation and indoor air quality.

Compared with experimental studies of indoor environment and building safety, CFD is less expensive and can obtain results much faster, due to the development in computing power and capacity as well as turbulence modeling. CFD can be applied to test flow and heat transfer conditions where experimental testing could prove very difficult, such as in space vehicles (Eckhardt and Zori 2002). Even if experimental measurements could be conducted, such an experiment would normally require hundreds of thousands dollars and many months of workers' time (Yuan et al. 1999).

However, CFD results cannot be always trusted, due to the assumptions used in turbulence modeling and approximations used in a simulation to simplify a complex real problem of indoor environment and building safety. Although a CFD simulation can always give a result for such a simulation, it may not necessarily give the correct result. A traditional approach to examine whether a CFD result is correct is by comparing the CFD result with corresponding experimental data. The question now is whether one can use a robust and validated CFD program, such as a well-known commercial CFD program, to solve a problem related to indoor environment and building safety without validation. This forms the main objective of the chapter.

This chapter presents a short review of the applications of CFD to indoor environment design and studies, and briefly introduces the most popular CFD models used. The chapter concludes that, although CFD is a powerful tool for indoor environment design and studies, a standard procedure must be followed so that the CFD program and user can be validated and the CFD results can be trusted. The procedure includes the use of simple cases that have basic flow features interested and experimental data available for validation. The simulation of indoor environment also requires creative thinking and the handling of complex boundary conditions. It is also necessary to play with the numerical grid resolution and distribution in order to get a grid-independent solution with reasonable computing effort. This investigation also discusses issues related to heat transfer. It is only through these incremental exercises that the user and the CFD program can produce results that can be trusted and used for indoor environment design and studies.

5.2 Computational fluid dynamics approaches

Indoor environment consists of four major components: thermal environment, indoor air quality, acoustics, and lighting environment. Building thermal environment and indoor air quality include the following parameters: air temperature, air velocity, relative humidity, environmental temperature, and contaminant and particulate concentrations, etc. The parameters concerning building safety are air temperature, smoke (contaminant and particulate) concentrations, flame temperature, etc. Obviously, normal CFD programs based on Navier–Stokes equations and heat and mass transfer cannot be used to solve acoustic and lighting components of an indoor environment.

However, the CFD programs can be used to deal with problems associated with thermal environment, indoor air quality, and building safety, since the parameters are solved by the programs. Hereafter, the chapter will use indoor environment to narrowly refer to thermal environment, indoor air quality, and building safety.

Almost all the flows in indoor environment are turbulent. Depending on how CFD solves the turbulent flows, it can be divided into direct numerical simulation, large eddy simulation (LES), and the Reynolds averaged Navier–Stokes equations with turbulence models (hereafter denotes as RANS modeling).

Direct numerical simulation computes turbulent flow by solving the highly reliable Navier–Stokes equation without approximations. Direct numerical simulation requires a very fine grid resolution to capture the smallest eddies in the turbulent flow at very small time steps, even for a steady-state flow. Direct numerical simulation would require a fast computer that currently does not exist and would take years of computing time for predicting indoor environment.

Large eddy simulation (Deardorff 1970) separates turbulent motion into large eddies and small eddies. This method computes the large eddies in a three-dimensional and time dependent way while it estimates the small eddies with a subgrid-scale model. When the grid size is sufficiently small, the impact of the subgrid-scale models on the flow motion is negligible. Furthermore, the subgrid-scale models tend to be universal because turbulent flow at a very small scale seems to be isotropic. Therefore, the subgrid-scale models of LES generally contain only one or no empirical coefficient. Since the flow information obtained from subgrid scales may not be as important as that from large scales, LES can be a general and accurate tool to study engineering flows (Lesieur and Metais 1996; Piomelli 1999). LES has been successfully applied to study airflow in and around buildings (Emmerich and McGrattan 1998; Murakami *et al.* 1999; Thomas and Williams 1999; Jiang and Chen 2002; Kato *et al.* 2003). Although LES requires a much smaller computer capacity and is much faster than direct numerical simulation, LES for predicting indoor environment demands a large computer capacity (10^{10} byte memory) and a long computing time (days to weeks).

The Reynolds averaged Navier–Stokes equations with turbulence models solve the statistically averaged Navier–Stokes equations by using turbulence transport models to simplify the calculation of the turbulence effect. The use of turbulence models leads to some errors, but can significantly reduce the requirement in computer memory and speed. The RANS modeling provides detailed information on indoor environment. The method has been successfully applied to the building indoor airflow and thermal comfort and indoor air quality analysis, as reviewed by Ladeinde and Nearon (1997) and Nielsen (1998). The RANS modeling can be easily used to study indoor environment. It would take only a few hours of computing time in a modern PC, should the RANS modeling be used to study a reasonable size of indoor environment.

In order to better illustrate the LES and RANS modeling, the following sections will discuss the fundamentals of the two CFD approaches. For simplicity, this chapter only discusses how the two CFD approaches solve Navier–Stokes equations and the continuity equation. Namely, the flow in indoor environment is considered to be isothermal and no gaseous and particulate contaminants and chemical reactions, are taken into account. In fact, temperature (energy), various contaminants, and various chemical reactions are solved in a similar manner.

5.2.1 Large-eddy simulation

By filtering the Navier–Stokes and continuity equations in the LES approach, one would obtain the governing equations for the large-eddy motions as

$$\frac{\partial \overline{u}_i}{\partial(t)} + \frac{\partial}{\partial x_j}(\overline{u_i u_j}) = -\frac{1}{\rho}\frac{\partial \overline{p}}{\partial x_i} + \nu\frac{\partial^2 \overline{u}_i}{\partial x_j \partial x_j} - \frac{\partial \tau_{ij}}{\partial x_j} \tag{5.1}$$

$$\frac{\partial \overline{u}_i}{\partial x_i} = 0 \tag{5.2}$$

where the bar represents grid filtering. The subgrid-scale Reynolds stresses, τ_{ij}, in Equation (5.1),

$$\tau_{ij} = \overline{u_i u_j} - \overline{u}_i \overline{u}_j \tag{5.3}$$

are unknown and must be modeled with a subgrid-scale model. Numerous subgrid-scale models have been developed in the past thirty years. The simplest and probably the most widely used is the Smagorinsky subgrid-scale model (Smagorinsky 1963) since the pioneering work by Deardorff (1970). The model assumes that the subgrid-scale Reynolds stress, τ_{ij}, is proportional to the strain rate tensor,

$$\overline{S}_{ij} = \frac{1}{2}\left(\frac{\partial \overline{u}_i}{\partial x_j} + \frac{\partial \overline{u}_j}{\partial x_i}\right) \tag{5.4}$$

$$\tau_{ij} = -2\nu_{SGS}\overline{S}_{ij} \tag{5.5}$$

where the subgrid-scale eddy viscosity, ν_{SGS}, is defined as

$$\nu_{SGS} = (C_{SGS}\Delta)^2(2\overline{S}_{ij}\cdot\overline{S}_{ij})^{1/2} = C\Delta^2(2\overline{S}_{ij}\cdot\overline{S}_{ij})^{1/2} \tag{5.6}$$

The Smagorinsky constant, C_{SGS}, ranges from 0.1 to 0.2 determined by flow types, and the model coefficient, C, is the square of C_{SGS}. The model is an adaptation of the mixing length model of RANS modeling to the subgrid-scale model of LES.

5.2.2 RANS modeling

Reynolds (1895) introduced the Reynolds-averaged approach in 1895. He decomposed the instantaneous velocity and pressure and other variables into a statistically averaged value (denoted with capital letters) and a turbulent fluctuation superimposed thereon (denoted with $'$ superscript). Taking velocity, pressure, and a scale variable as examples:

$$u_i = U_i + u_i', \quad p = P + p', \quad \phi = \Phi + \phi' \tag{5.7}$$

The statistical average operation on the instantaneous, averaged, and fluctuant variables have followed the Reynolds average rules. Taking velocity as an example, the Reynolds average rules can be summarized as:

$$\overline{u_i} = \overline{U_i} = U_i, \quad \overline{u_i'} = 0, \quad \overline{u_i'U_j} = 0,$$
$$\overline{u_i + u_j} = U_i + U_j, \quad \overline{u_i u_j} = U_i U_j + \overline{u_i' u_j'}$$

(5.8)

Note that the bars in Equation 5.8 stand for "statistical average" and are different from those used for LES. In LES, those bars represent grid filtering.

By applying the Reynolds averaging method to the Navier–Stokes and continuity equation, they become:

$$\frac{\partial U_i}{\partial t} + U_j \frac{\partial U_i}{\partial x_j} = -\frac{1}{\rho} \frac{\partial P}{\partial x_i} + \frac{\partial}{\partial x_j} \left(\nu \frac{\partial U_i}{\partial x_j} - \overline{u_i' u_j'} \right)$$

(5.9)

$$\frac{\partial U_i}{\partial x_i} = \frac{\partial u_i'}{\partial x_i} = 0$$

(5.10)

where $\overline{u_i' u_j'}$ is the Reynolds stress that is unknown and must be modeled. In the last century, numerous turbulence models have been developed to represent $\overline{u_i' u_j'}$. Depending on how the Reynolds stress is modeled, RANS turbulence modeling can be further divided into Reynolds stress models and eddy-viscosity models. For simplicity, this chapter discusses only eddy-viscosity turbulence models that adopt the Boussinesq approximation (1877) to relate Reynolds stress to the rate of mean stream through an "eddy" viscosity ν_t.

$$\overline{u_i' u_j'} = \frac{2}{3} \delta_{ij} k - \nu_t \left(\frac{\partial U_i}{\partial x_j} + \frac{\partial U_j}{\partial x_i} \right)$$

(5.11)

where δ_{ij} is the Kronecker delta (when $i \neq j$, $\delta_{ij} = 0$; and when $i = j$, $\delta_{ij} = 1$), and k is the turbulence kinetic energy ($k = \overline{u_i' u_i'}/2$). Among hundreds of eddy-viscosity models, the standard k–ε model (Launder and Spalding 1974) is most popular. The standard k–ε model solves eddy viscosity through

$$\nu_t = C_\mu \frac{k^2}{\varepsilon}$$

(5.12)

where $C_\mu = 0.09$ is an empirical constant. The k and ε can be determined by solving two additional transport equations:

$$U_j \frac{\partial k}{\partial x_j} = \frac{\partial}{\partial x_j} \left[\left(\nu + \frac{\nu_t}{\sigma_k} \right) \frac{\partial k}{\partial x_j} \right] + P - \varepsilon$$

(5.13)

$$U_j \frac{\partial \varepsilon}{\partial x_j} = \frac{\partial}{\partial x_j} \left[\left(\nu + \frac{\nu_t}{\sigma_\varepsilon} \right) \frac{\partial \varepsilon}{\partial x_j} \right] + [C_{\varepsilon 1} P - C_{\varepsilon 2} \varepsilon] \frac{\varepsilon}{k}$$

(5.14)

where

$$P = \nu_t \frac{1}{2} \left(\frac{\partial U_i}{\partial x_j} + \frac{\partial U_j}{\partial x_i} \right)^2$$

(5.15)

and $\sigma_k = 1.0$, $\sigma_\varepsilon = 1.3$, $C_{\varepsilon 1} = 1.44$, and $C_{\varepsilon 2} = 1.92$ are empirical constants. The two-equation k–ε model is most popular but not the simplest one. The simplest ones are zero-equation turbulence models, such as the constant viscosity model and the one proposed by Chen and Xu (1998). The constant viscosity model and zero-equation models do not solve turbulence quantities by transport equations.

Be it LES or RANS modeling, the abovementioned equations cannot be solved analytically because they are highly nonlinear and interrelated. However, they can be solved numerically on a computer by discretizing them properly with an appropriate algorithm. Many textbooks have been devoted to this topic. Due to limited space available, this chapter does not discuss this issue here. Finally, boundary conditions must be specified in order to make the equations solvable for a specific problem of indoor environment.

If one has used a CFD program with the abovementioned equations and specified boundary conditions for a flow problem, can one trust the results obtained? The following section will use an example to illustrate how one could obtain CFD results for an indoor environment problem and how one could evaluate the correctness of the results.

5.3   Simulation and analysis

The following example is a study of indoor air and contaminant distribution in a room with displacement ventilation, as shown in Figure 5.1. The room was 5.16 m long, 3.65 m wide, and 2.43 m high. Cold air was supplied through a diffuser in the lower part of a room, and warm air was exhausted at the ceiling level. The two-person office contained many heated and unheated objects, such as occupants, lighting, computers, and furniture. For this case, Yuan *et al.* (1999) measured the air temperature, air velocity, and contaminant concentration by using SF_6 as a tracer-gas. The tracer-gas was used to simulate contaminant emissions from the two occupants, such as CO_2. The temperature of the inlet airflow from the diffuser was 17.0°C and the ventilation rate was 183 m³/h. The total heat sources in the room were 636 W.

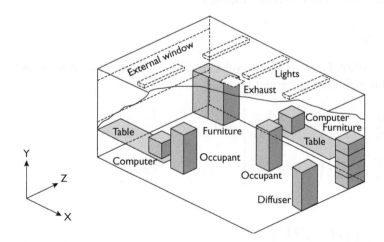

Figure 5.1 The schematic of a room with mixed convection flow.

5.3.1 General problems in using CFD programs

This is a project the author assigned to train his graduate students in gaining experience and confidence in using a well-validated commercial CFD program. The graduate students majored in mechanical engineering and had sufficient knowledge of fluid dynamics, heat transfer, and numerical methods. Without exception, no student could obtain correct results in the first instance when they attempted to directly solve such a problem. Their CFD results were compared with the experimental data from Yuan *et al.* (1999). The problems can be summarized as follows:

- difficulty in selecting a suitable turbulence model;
- incorrect setting of boundary conditions for the air-supply diffuser;
- inappropriate selection of grid resolution;
- failure to estimate correctly convective portion of the heat from the heat sources, such as the occupants, computers, and lighting;
- improper use of numeric techniques, such as relaxation factors and internal iteration numbers.

For such a problem as shown in Figure 5.1, both the LES and RANS approaches were suitable. Through the RANS approach, many commercial CFD programs offer numerous turbulence models for CFD users. It is a very challenging job for a beginner to decide which model to use. Although for some cases, more sophisticated models can generate more accurate results, our experience found that the Smagorinsky subgrid-scale model for LES and the standard $k–\varepsilon$ model for RANS are more universal, consistent, and stable. Unfortunately, they do not always produce accurate results and can perform poorer than other models in some cases.

Simulation of a specific problem of indoor environment requires creative approaches. One typical example is how to simulate the air-supply diffuser, which is a perforated panel with an effective area of less than 10%. Some commercial codes have a library of diffusers that can be used to simulate an array of complex diffusers, such as Airpak from Fluent. Without such a library, we found that only experienced CFD users may know how to simulate such a diffuser.

Since the geometry of the displacement ventilation case is rectangular, many of the students would select a grid distribution that fits the boundaries of the objects in the room. The grid size would be selected in such a way that no interpolation is needed to obtain results in places of interest. Not everyone would refine the grid resolution to obtain grid-independent results. It is hard to obtain grid-independent results, especially when LES is used. When a wall-function is used for boundary layers, it is very rare that a CFD user would check if the grid resolution near a wall is satisfactory.

ASHRAE (Chen and Srebric 2002) has developed a guide on using CFD to simulate indoor environment. One major emphasis is on establishing a CFD model that could simulate a specific problem. If we take the displacement ventilation case as an example, it is not an easy task to provide the thermal and fluid boundary conditions. For example, it is difficult to estimate temperature or heat fluxes for the building enclosure surfaces and the heated objects, such as computers, occupants, and lighting. As a consequence, the mean air temperature computed by different users with the same CFD program can differ as much as 3 K.

Most CFD programs, especially the commercial ones, are generalized and designed to solve flow and heat and mass transfer, not just for simulating indoor environment. As a result, the CFD programs provide many options. A user can fine-tune the parameters to obtain a result. The parameters that can be tuned include, but are not limited to, model coefficients, relaxation factors, and iteration numbers. With different tuning values, the CFD results are often not the same.

Therefore, a CFD beginner, who attempted to solve flow and heat and mass transfer for the displacement ventilation case, became frustrated when he/she found that his/her CFD results were different from the measured data. If no measured data were available for comparison, the user would have no confidence about the correctness of the CFD results. In order to correctly perform a CFD simulation for a specific flow problem related to indoor environment, we strongly recommend the use of ASHRAE procedure for verification, validation, and reporting of indoor environment CFD analyses (Chen and Srebric 2002).

5.3.2   How to conduct CFD analyses of indoor environment

To design or study an indoor environment problem with CFD, one needs to

- confirm the abilities of the turbulence model and other auxiliary models to predict all physical phenomena in the indoor environment;
- confirm the discretization method, grid resolution, and numerical algorithm for the flow simulation;
- confirm the user's ability to use the CFD code to perform indoor environment analyses.

The confirmations are indeed a validation process through which a user can know his/her ability to perform a CFD simulation and the correctness of the CFD results. If the user is asked to simulate the displacement ventilation case, no experimental data is available for comparison, as in most indoor environment designs and studies. The validation would use several subsystems that represent the complete flow, heat and mass transfer features of the case. For the displacement ventilation that has a mixed convection flow, the user may start a two-dimensional natural convection in a cavity and a forced convection in a cavity. Since mixed convection is a combination of natural and forced convection, the two subsystems can represent the basic flow features of the displacement ventilation. Of course, CFD validation is not only for flow type; the CFD validation should be done in progressive stages. A typical procedure for correctly simulating the displacement ventilation would be as follows:

- Simulation of a two-dimensional natural convection case.
- Simulation of a two-dimensional forced convection case.
- Simulation of a simple three-dimensional case.
- Simulation of complex flow components.
- Change in grid resolution, especially the resolution near walls.
- Calculation of convective/radiative ratio for different heat sources.
- Simulation of the displacement ventilation.

This procedure is incremental in the complexity of the CFD simulations. Since it is relatively easy to judge the correctness of the CFD results for simple cases (many of them have experimental data available in literature), the user can gain confidence in the simulation exercise. While such a simulation seems to take longer time than direct simulation of the displacement ventilation, the procedure is more effective and can actually obtain the correct results for the displacement ventilation, rather than directly solving the case without the basic exercise. This is because the CFD user would have a hard time to find out where the simulation has gone wrong, due to the complexity of the displacement ventilation and inexperience in usage of the CFD program. The following sections illustrate the simulation procedure.

5.3.3 Simulation of a two-dimensional natural convection case

The two-dimensional natural convection case concerns flow in a cavity of 0.5 m width and 2.5 m height, as shown in Figure 5.2. Cheesewright *et al.* (1986) conducted the experimental studies on this case. The experiment maintained isothermal conditions (64.8°C and 20°C) on the two vertical walls and insulated the two horizontal walls, even though they were not ideally insulated. The Rayleigh number (Ra) based on the cavity height (h) was 5×10^5. The simulation employed both the zero-equation model (Chen and Xu 1998) and the standard k–ε model.

Figure 5.3(a) compares the computed and measured mean velocity at the mid-height of the cavity, which shows good agreement except at the near-wall regions. The standard k–ε model with the wall function appears to capture the airflows near the surfaces better than the zero-equation model. The predicted core air temperatures with the k–ε model, as shown in Figure 5.3(b), also agree well with Cheesewright's measurements. The results with the zero-equation model are higher

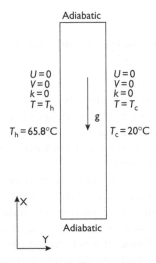

Figure 5.2 Geometry and boundary conditions for two-dimensional natural convection in a cavity.

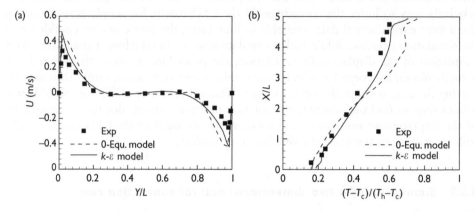

Figure 5.3 (a) The vertical velocity profile at mid-height and (b) temperature profile in the mid-height for two-dimensional natural convection case.

than the measurements, although the computed and measured temperature gradients in the core region are similar. A beginner may not be able to find the reasons for the discrepancies. With the use of two models, it is possible to find that different models do produce different results.

Since displacement ventilation consists of natural and forced convection, it is necessary to simulate a forced convection in order to assess the performance of the turbulence models. A case proposed by Nielsen (1974) with experimental data is most appropriate. Due to limited space available, this chapter does not report the simulation results. In fact, the zero-equation model and the k–ε model have performed similarly for the two-dimensional forced convection case as they did for the natural convection case reported earlier.

5.3.4   Simulation of a three-dimensional case without internal obstacles

The next step is to simulate a three-dimensional flow. As the problem becomes more complicated, the experimental data often becomes less detailed and less reliable in terms of quality. Fortunately, with the experience of the two-dimensional flow simulation, the three-dimensional case selection is not critical. For example, the experimental data of mixed convection in a room as shown in Figure 5.4 from Fisher (1995) seems appropriate for this investigation.

Figure 5.5 presents the measured and calculated air speed contours, which show the similarity between the measurement and simulation of the primary airflow structures. The results show that the jet dropped down to the floor of the room after traveling forward for a certain distance due to the negative buoyancy effect. This comparison is not as detailed quantitatively as the two-dimensional natural convection case. However, a CFD user would gain some confidence in his/her results through this three-dimensional simulation.

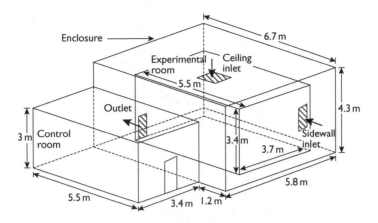

Figure 5.4 Schematic of experimental facility (Fisher 1995).

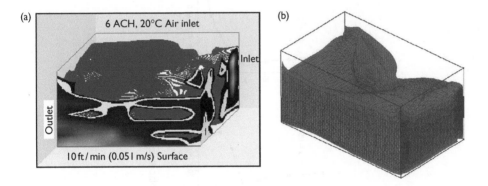

Figure 5.5 Air speed contour in the room (a) as measured; (b) as simulated by CFD. (See Plate II.)

5.3.5 Simulation of complex flow components

A room normally consists of several complex flow elements, such as air-supply diffusers, irregular heat sources, and complicated geometry. Correct modeling of these flow components is essential for achieving accurate simulation of airflow in the room. This chapter takes an air-supply diffuser used for displacement ventilation as an example for illustrating how the complex flow components should be modeled.

Figure 5.6 shows the flow development in front of a displacement diffuser. The jet drops immediately to the floor in the front of the diffuser because of the low air supply velocity and buoyancy effect. The jet then spreads over the floor and reaches the opposite wall. In front of the diffuser, the jet velocity profile changes along its trajectory. Close to the diffuser, no jet formula can be used since the jet is in a transition region. Only after 0.9 m (3.0 ft) does the jet form an attached jet, where a jet formula could be used. However, jet formulae can only predict velocities in the jet region that is less than 0.2 m above the floor, because the velocities above the region are influenced

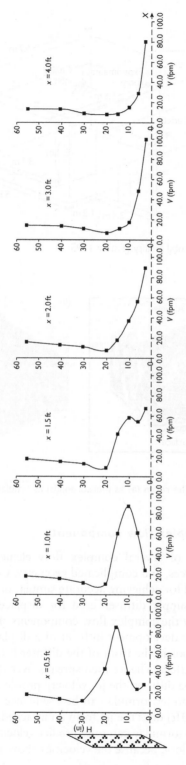

Figure 5.6 Development of the wall jet in front of the displacement diffuser.

by the room conditions. In fact, the velocity profile above the jet region represents the backward airflow towards the displacement diffuser.

Chen and Moser (1991) proposed a momentum method that decouples momentum and mass boundary conditions for the diffuser in CFD simulation. The diffuser is represented in the CFD study with an opening that has the same gross area, mass flux, and momentum flux as a real diffuser does. This model enables specification of the source terms in the conservation equations over the real diffuser area. The air supply velocity for the momentum source term is calculated from the mass flow rate, $\dot{m}$, and the diffuser effective area A_0:

$$U_0 = \dot{m}/(\rho\,A_0) \tag{5.16}$$

Srebric (2000) demonstrated that the momentum method can produce satisfactory results, and the method is thus used for this investigation. As one can see, modeling of a complex flow element requires substantial effort and knowledge.

5.3.6 Change in grid resolution, especially the resolution near walls

So far we have discussed the establishment of a CFD model for displacement ventilation. Numerical procedure is equally important in achieving accurate results. In most cases, one would demand a grid-independent solution. By using Fisher's case (1995) as an example, this investigation has used four sets of grids to simulate the indoor airflow: a coarse grid ($22 \times 17 \times 15 = 5{,}610$ cells), a moderate grid ($44 \times 34 \times 30 = 44{,}880$ cells), a fine grid ($66 \times 51 \times 45 = 151{,}470$ cells), and a locally refined coarse grid ($27 \times 19 \times 17 = 8{,}721$ cells) that has the same resolution in the near-wall regions as the fine grid.

Figure 5.7 presents the predicted temperature gradient along the vertical central line of the room with the different grid resolutions. Obviously, a coarse grid distribution cannot produce satisfactory results. The moderate and fine grid systems produced similar temperature profile and could be considered as grid independent.

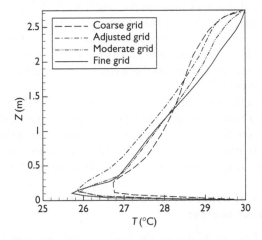

Figure 5.7 Predicted temperature gradient along the vertical central line of the room.

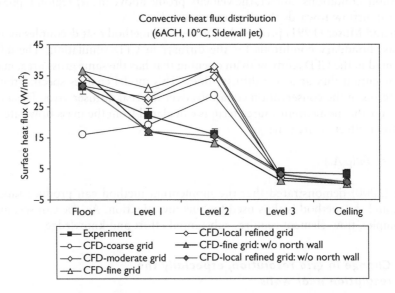

Figure 5.8 Comparison of convective heat fluxes from enclosures with various grid resolutions. (See Plate III.)

It is also interesting to know that by using locally refined grid distribution, a coarse grid system can yield satisfactory results.

The grid distribution has a significant impact on the heat transfer. Figure 5.8 shows the predicted convective heat fluxes from enclosures with different grid systems. The convective heat fluxes from the floor predicted with the refined grid systems are much closer to the measurement than those with the coarse grid. However, the difference between the measured and simulated results at wall Level 2 is still distinct, even with the fine grid. The analysis indicates that the impact of the high-speed jet flow on Level 2 of the north wall is the main reason for the large heat flux at the entire wall Level 2. Since the vertical jet slot is very close to the north wall, the cold airflow from the jet inlet causes the strong shear flow at the north wall, introducing the extra heat transfer at this particular area. The experiment did not measure this heat transfer zone within the inner jet flow. If the north wall was removed from the analysis of the wall convective heat fluxes, the agreement between the computed results and measured data would be much better.

Figure 5.8 also indicates that, instead of using a global refined grid that may need long computing time, a locally refined coarse grid can effectively predict the airflow and heat transfer for such an indoor case. Good resolution for the near-wall regions is much more important than for the inner space because the air temperature in the core of a space is generally more uniform than that in the perimeter of a space.

5.3.7 Calculation of convective/radiative ratio for different heat sources

In most indoor airflow simulations, the building interior surface temperatures are specified as boundary conditions. Then, the heat from heat sources must be split into convective and radiative parts. The convective part is needed as boundary conditions for the CFD simulation, while the radiative part is lumped into the wall surface temperatures. This split can be rather difficult, since the surface temperature of the heat sources and/or the surface area are unknown in most cases. Without a correct split, the final air temperature of the room could deviate a few degrees from the correct one. Therefore, the split would require a good knowledge of heat transfer. This problem will not be discussed in detail here, since it is problem dependent. For the displacement ventilation, the convective/radiative ratio should be 80/20 for occupants, 56/44 for computers, and 60/40 for lighting.

5.3.8 Simulation of displacement ventilation

With all the exercises given earlier, a CFD user would gain sufficient experience in indoor environment simulation by CFD. The user could use CFD to study indoor environment, such as airflow in a room with displacement ventilation (as shown in Figure 5.1), with confidence. The results will then be somewhat trusted.

This section shows the CFD results computed by the coauthor for the displacement ventilation case (Figure 5.1). The experimental data from Yuan *et al.* (1999) was available for this case. The data is used as a comparison in showing if the CFD results can be trusted.

This investigation used a CFD program with the zero-equation turbulence model and the standard $k-\varepsilon$ model. The computational grid is $55 \times 37 \times 29$, which is sufficient for obtaining the grid-independent solution, according to Srebric (2000) and our experience in Fisher's case (1995). Figure 5.9(a) shows the calculated air velocity and

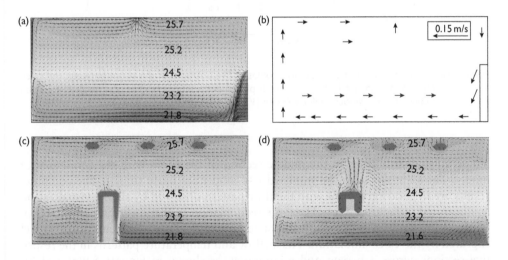

Figure 5.9 Velocity and temperature distributions for the displacement ventilation case (a) calculated results in the middle section, (b) observed airflow pattern with smoke visualization in the middle section, (c) calculated results in the section across a computer, and (d) calculated results in the section across an occupant. (See Plate IV.)

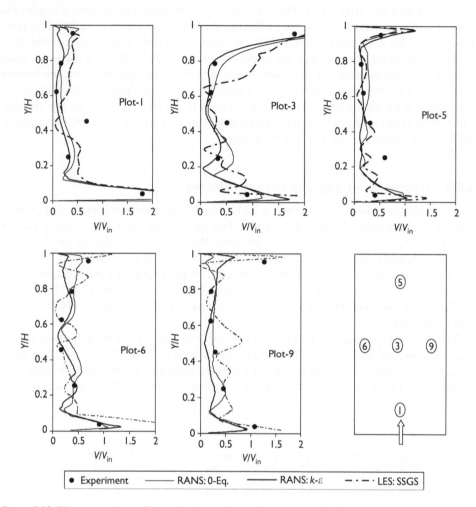

Figure 5.10 The comparison of the velocity profiles at five positions in the room between the calculated and measured data for the displacement ventilation case. $Z =$ height/total room height (H), $V =$ velocity/inlet velocity (V_{in}), $H = 2.43$ m, $V_{in} = 0.086$ m/s.

temperature distributions in the middle section of the room with the zero-equation model. The solutions with the standard k–ε model are fairly similar. The computed results are in good agreement with the flow pattern observed by smoke visualization, as illustrated in Figure 5.9(b). The large recirculation in the lower part of the room, which is known as a typical flow characteristic of displacement ventilation, is well captured by the CFD simulation. The airflow and temperature patterns in the respective sections across a person and a computer, as shown in Figures 5.9(c) and (d), clearly exhibit the upward thermal plumes due to the positive buoyancy from the heat sources.

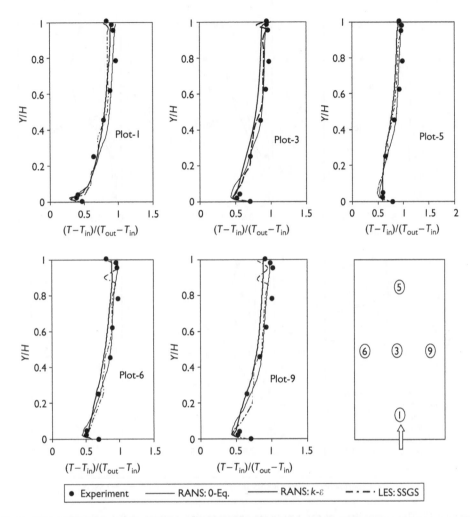

Figure 5.11 The comparison of the temperature profiles at five positions in the room between the calculated and measured data for the displacement ventilation case. $Z =$ height/total room height (H), $T = (T_{air} - T_{in}/T_{out} - T_{in})$, $H = 2.43$ m, $T_{in} = 17.0°C$, $T_{out} = 26.7°C$.

The study further compared the measured and calculated velocity, air temperature, and tracer-gas concentration (SF_6 used to simulate bio-effluent from the two occupants) profiles at five locations where detailed measurements were carried out. The locations in the floor plan are illustrated in the lower-right of Figures 5.10–5.12. The figures show the computed results by RANS modeling with the zero-equation model and the standard k–ε model, and large-eddy simulation with the Smogrinsky subgrid-scale (SSGS) model.

Clearly, the computed results are not exactly the same as the experimental data. In fact, the two results will never be the same due to the approximations used in

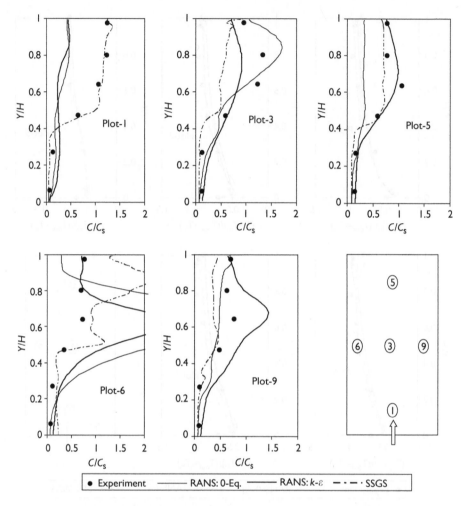

Figure 5.12 The comparison of the tracer-gas concentration profiles at five positions in the room between the calculated and measured data for the displacement ventilation case. Z = height/total room height (H), $H = 2.43\,\text{m}$, $C_s = 0.42\,\text{ppm}$.

CFD and errors in the measuring equipment and experimental rig. The agreement is better for temperature than the velocity and tracer-gas concentration. Since omnidirectional anemometers were used to measure air velocity and the air velocity is low, the convection caused by probes would generate a false velocity of the same magnitude. Therefore, the accuracy of the measured velocity is not very high. For tracer-gas concentration, the airflow pattern is not very stable and measuring SF_6 concentration at a single point would take 30 s. The measurement has a great uncertainty as well.

On the other hand, the performance of the CFD models is also different. The LES results seem slightly better than the others. Since LES uses at least one-order magnitude computing time than the RANS modeling, LES seems not worth in such an application. The profile curves are not very smooth that may indicate more averaging time needed.

Nevertheless, the CFD results do reproduce the most important features of airflow in the room, and can quantitatively predict the air distribution. The discrepancies between the computed results and experimental data can be accepted for indoor environment design and study. We may conclude that the CFD results could be trusted for this case even if no experimental data were available for validation.

5.4 Conclusions

This chapter shows that applications of CFD program to indoor environment design and studies need some type of validation of the CFD results. The validation is not only for the CFD program but also for the user. The validation process will be incremental, since it is very difficult to obtain correct results for a complex flow problem in indoor environment.

This chapter demonstrated the validation procedure by using displacement ventilation in a room as an example. The procedure suggests using two-dimensional cases for selecting a turbulence model and employing an appropriate diffuser model for simplifying complex flow components in the room, such as a diffuser. This chapter also demonstrates the importance in performing grid-independent studies and other technical issues. With the exercises, one would be able to use a CFD program to simulate airflow distribution in a room with displacement ventilation, and the CFD results can be trusted.

5.5 Acknowledgment

This investigation is supported by the United States National Institute of Occupational, Safety, and Health (NIOSH) through research grant No. 1 R01 OH004076-01.

Nomenclature

A_0	Effective area of a diffuser	p	air pressure (Pa)
C	Smagorinsky model coefficient	S_{ij}	strain rate tensor (1/s)
C_{SGS}	Smagorinsky model constant	t	time (s)
$C_{\varepsilon 1}$	coefficient in k–ε model	U_i, U_j	averaged air velocity components
$C_{\varepsilon 2}$	coefficient in k–ε model		in the x_i and x_j directions (m/s)
C_μ	coefficient in k–ε model	U_0	face velocity at a diffuser
k	kinetic energy (J/kg)	u_i, u_j	air velocity components in the x_i
P	averaged air pressure (Pa)		and x_j directions (m/s)
$\dot{m}$	Mass flow rate (kg/s)	x_i, x_j	coordinates in i and j directions (m)

Greek symbols

Δ	filter width (m)	ρ	air density (kg/m^3)
δ	Kronecker delta	σ_k	Prandlt number for k

ε	dissipation rate of kinetic energy (W/kg)	σ_ε	Prandlt number for ε
ν	air kinematic viscosity (m²/s)	τ_{ij}	subgrid-scale Reynolds stresses (m²/s²)
ν_{SGS}	subgrid-scale eddy viscosity (m²/s)		
ν_t	turbulent air kinematic viscosity (m²/s)	ϕ	scalar variables
		Φ	averaged scalar variables

Superscripts

− grid filtering or Reynolds averaging

′ fluctuating component of a variable

References

Bartak, M., Beausoleil-Morrison, I., Clarke, J.A., Denev, J., Drkal, F., Lain, M., Macdonald, I.A., Melikov, A., Popiolek, Z., and Stankov, P. (2002). "Integrating CFD and building simulation." *Building and Environment*, Vol. 37, No. 8, pp. 865–871.

Beausoleil-Morrison, I. (2002). "The adaptive conflation of computational fluid dynamics with whole-building thermal simulation." *Energy and Buildings*, Vol. 34, No. 9, pp. 857–871.

Boussinesq, J. (1877). "Théorie de l'écoulement tourbillant." *Mem. Présentés par Divers Savants Acad. Sci. Inst. Fr*, Vol. 23, pp. 46–50.

Carrilho da Graca, G., Chen, Q., Glicksman, L.R., and Norford, L.K. (2002). "Simulation of wind-driven ventilative cooling systems for an apartment building in Beijing and Shanghai." *Energy and Buildings*, Vol. 34, No. 1, pp. 1–11.

Cheesewright, R., King, K.J., and Ziai, S. (1986). "Experimental data for the validation of computer codes for the prediction of two-dimensional buoyant cavity flows." In J.A.C. Humphrey, C.T. Adedisian, and B.W. le Tourneau (eds) *Significant Questions in Buoyancy Affected Enclosure or Cavity Flows*, ASME, pp. 75–81.

Chen, Q. and Moser, A. (1991). "Simulation of a multiple-nozzle diffuser." In *Proceedings of the 12th AIVC Conference on Air Movement and Ventilation Control within Buildings*. Ottawa, Canada, Vol. 2, pp. 1–13.

Chen, Q. and Xu, W. (1998). "A zero-equation turbulence model for indoor airflow simulation." *Energy and Buildings*, Vol. 28, No. 2, pp. 137–144.

Chen, Q. and Srebric, J. (2002). "A procedure for verification, validation, and reporting of indoor environment CFD analyses." *International Journal of HVAC&R Research*, Vol. 8, No. 2, pp. 201–216.

Deardorff, J.W. (1970). "A numerical study of three-dimensional turbulent channel flow at large Reynolds numbers." *Journal of Fluid Mechanics*, Vol. 41, pp. 453–480.

Eckhardt, B. and Zori, L. (2002). "Computer simulation helps keep down costs for NASA's 'lifeboat' for the international space station." *Aircraft Engineering and Aerospace Technology: An International Journal*, Vol. 74, No. 5, pp. 442–446.

Emmerich, S.J. and McGrattan, K.B. (1998). "Application of a large eddy simulation model to study room airflow." *ASHRAE Transactions*, Vol. 104, pp. 1128–1140.

Fisher, D.E. (1995). "An experimental investigation of mixed convection heat transfer in rectangular enclosure." PhD thesis. University of Illinois at Urbana-Champaign, Illinois.

Fisk, W.J. (2000). "Health and productivity gains from better indoor environments and their relationship with energy efficiency." *Annual Review of Energy and the Environment*, Vol. 25, pp. 537–566.

Huang, H. and Haghighat, F. (2002). "Modelling of volatile organic compounds emission from dry building materials." *Building and Environment*, Vol. 37, No. 12, pp. 1349–1360.

Jiang, Y. and Chen, Q. (2002). "Effect of fluctuating wind direction on cross natural ventilation in building from large eddy simulation." *Building and Environment*, Vol. 37, No. 4, pp. 379–386.

Kato, S., Ito, K., and Murakami, S. (2003). "Analysis of visitation frequency through particle tracking method based on LES and model experiment." *Indoor Air*, Vol. 13, No. 2, pp. 182–193.

Ladeinde, F. and Nearon, M. (1997). "CFD applications in the HVAC&R industry." *ASHRAE Journal*, Vol. 39, No. 1, p. 44.

Launder, B.E. and Spalding, D.B. (1974). "The numerical computation of turbulent flows." *Computer Methods in Applied Mechanics and Energy*, Vol. 3, pp. 269–289.

Lesieur, M. and Metais, O. (1996). "New trends in large eddy simulations of turbulence." *Annual Review of Fluid Mechanics*, Vol. 28, pp. 45–82.

Lo, S.M., Yuen, K.K., Lu, W.Z., and Chen, D.H. (2002). "A CFD study of buoyancy effects on smoke spread in a refuge floor of a high-rise building." *Journal of Fire Sciences*, Vol. 20, No. 6, pp. 439–463.

Manz, H. (2003). "Numerical simulation of heat transfer by natural convection in cavities of facade elements." *Energy and Buildings*, Vol. 35, No. 3, pp. 305–311.

Murakami, S., Iizuka, S., and Ooka, R. (1999). "CFD analysis of turbulent flow past square cylinder using dynamic LES." *Journal of Fluids and Structures*, Vol. 13, No. 78, pp. 1097–1112.

Murakami, S., Kato, S., Ito, K., and Zhu, Q. (2003). "Modeling and CFD prediction for diffusion and adsorption within room with various adsorption isotherms." *Indoor Air*, Vol. 13, No. 6, pp. 20–27.

Nielson, P.V. (1974). "Flow in air conditioned rooms." PhD thesis. Technical University of Denmark, Copenhagen.

Nielsen, P.V. (1998). "The selection of turbulence models for prediction of room airflow." *ASHRAE Transactions*, Vol. 104, No. 1, pp. 1119–1127.

Piomelli, U. (1999). "Large eddy simulation: achievements and challenges." *Progress in Aerospace Sciences*, Vol. 35, pp. 335–362.

Quinn, A.D., Wilson, M., Reynolds, A.M., Couling, S.B., and Hoxey, R.P. (2001). "Modelling the dispersion of aerial pollutants from agricultural buildings—an evaluation of computational fluid dynamics (CFD)." *Computers and Electronics in Agriculture*, Vol. 30, No. 1, pp. 219–235.

Reynolds, O. (1895). "On the dynamical theory of incompressible viscous fluids and the determination of the criterion." *Philosophical Transactions of the Royal Society of London, Series A*, Vol. 186, p. 123.

Sahm, P., Louka, P., Ketzel, M., Guilloteau, E., and Sini, J.-F. (2002). "Intercomparison of numerical urban dispersion models—part I: street canyon and single building configurations." *Water, Air and Soil Pollution: Focus*, Vol. 2, Nos (5–6), pp. 587–601.

Smagorinsky, J. (1963). "General circulation experiments with the primitive equations. I. The basic experiment." *Monthly Weather Review*, Vol. 91, pp. 99–164.

Srebric, J. (2000). "Simplified methodology for indoor environment design." PhD thesis. Department of Architecture, Massachusetts Institute of Technology, Cambridge, MA.

Swanddiwudhipong, S. and Khan, M.S. (2002). "Dynamic response of wind-excited building using CFD." *Journal of Sound and Vibration*, Vol. 253, No. 4, pp. 735–754.

Thomas, T.G. and Williams, J.J.R. (1999). "Generating a wind environment for large eddy simulation of bluff body flows—a critical review of the technique." *Journal of Wind Engineering and Industrial Aerodynamics*, Vol. 82, No. 1, pp. 189–208.

Topp, C., Nielsen, P.V., and Heiselberg, P. (2001). "Influence of local airflow on the pollutant emission from indoor building surfaces." *Indoor Air*, Vol. 11, No. 3, pp. 162–170.

Tsou J.-Y. (2001). "Strategy on applying computational fluid dynamic for building performance evaluation." *Automation in Construction*, Vol. 10, No. 3, pp. 327–335.

Woods, J.E. (1989). "Cost avoidance and productivity in owning and operating buildings." *Occupational Medicine: State of the Art Reviews*, Vol. 4, No. 4, pp. 753–770.

Yeoh, G.H., Yuen, R.K.K., Lo, S.M., and Chen, D.H. (2003). "On numerical comparison of enclosure fire in a multi-compartment building." *Fire Safety Journal*, Vol. 38, No. 1, pp. 85–94.

Yuan, X., Chen, Q., Glicksman, L.R., Hu, Y., and Yang, X. (1999). "Measurements and computations of room airflow with displacement ventilation." *ASHRAE Transactions*, Vol. 105, No. 1, pp. 340–352.

Zhai, Z. and Chen, Q. (2003). "Solution characters of iterative coupling between energy simulation and CFD programs." *Energy and Buildings*, Vol. 35, No. 5, pp. 493–505.

Chapter 6

New perspectives on Computational Fluid Dynamics simulation

D. Michelle Addington

6.1 Introduction

Simulation modeling, particularly Computational Fluid Dynamics (CFD), has opened an unprecedented window into understanding the behavior of building environments. The late entry of these tools into the building arena—more than 20 years after their initial application in the aerospace industry—is indicative of the complexity of building air behavior. Unlike many applications, such as turbomachinery or nuclear power cooling, in which one or two mechanisms may dominate, building air flow is a true mixing pot of behaviors: wideranging velocities, temperature/density stratifications, transient indoor and outdoor conditions, laminar and turbulent flows, conductive, convective and radiant transfer, and random heat and/or mass generating sources. As impossible to visualize as it was to determine, building air behavior represented one of the last problems in classical physics to be understood. CFD offers system designers as well as architects and building owners a "picture" of the complex flow patterns, finally enabling an escape from the all too often generically designed system.

Nevertheless, the true potentials of CFD and simulation modeling have yet to be exploited for building applications. In most other fields, including automotive design, aeronautics, and electronics packaging, CFD has been used for much more than just a test and visualization tool for evaluating a specific installation or technology. Rather, many consider CFD to be a fundamental means of describing the basic physics, and, as such, its numerical description completes the triad with analytical and empirical descriptions. Given that the major technology for heating and cooling buildings (the HVAC system) has been in place for nearly a century with only minor changes, a large opportunity could be explored if CFD were used to characterize and understand the physical phenomena taking place in a building, possibly even leading to a challenge of the accepted standard of the HVAC system.

This chapter will address the differences between building system modeling and phenomenological modeling: the relative scales, the different modeling requirements, the boundaries, and modes of evaluation. In particular, the chapter will examine the fundamental issues raised when density-driven convection is treated as the primary mode of heat transfer occurring in buildings. This is in contrast to the more normative privileging of the pressure-driven or forced convection produced from HVAC systems that is more often the focus of CFD simulations in building. The chapter will then conclude with a discussion of how the exploration of phenomenological behaviors could potentially lead to the development of unprecedented technological responses.

Unlike many other computational tools intended for building optimization, CFD simulation modeling allows for discrete prediction of transient conditions. Prior to the introduction of these tools, the behavior of building environments was often described anecdotally or determined from exhaustive physical data. The tendency for anecdotal descriptions began in the nineteenth century when engineers and scientists used arrows to elaborately diagram air movement due to convection, suggesting physically impossible paths. Their diagrams were so convincing that the governments of Great Britain and the United States devoted substantial funds to nonsensical and eventually ill-fated modifications of the Houses of Parliament and the US Capitol, all intended to improve the air movement inside the buildings (Elliot 1992). Many of the current descriptive models, however, are no less anecdotal, often replicating common assumptions about how air moves, without recognition of its markedly nonintuitive behavior. For example, double-skin facades have routinely been described and justified as providing a greenhouse-like effect in the winter and a thermal chimney effect in the summer, but such generic assumptions have little in common with the very complex behavior of this multiply layered system sandwiched between transient air masses.

More than simply a tool for visualization, CFD provides a method for "solving" the Navier–Stokes equations—the fundamental equations governing heat and mass transfer—whose complex nonlinearity had rendered them all but impossible to solve until the Cray-1 supercomputer was developed approximately three decades ago. The use of CFD has revolutionized many disciplines from aeronautics to nuclear engineering, and its impact has been felt throughout the microelectronics industry (today's fast processors are feasible primarily because of the innovative heat shedding strategies made possible through CFD analysis). As simplified variations with user-friendly interfaces became more readily available, CFD began to penetrate the field of building systems analysis over the last decade. The initial applications, however, were quite unsophisticated in comparison to contemporary investigations in other fields. As an example, monographs on the Kansai airport highlight the visualizations produced by an early CFD code, and the common lore is that the results were used to determine the optimum shape of the terminal's roof, notwithstanding that the grid size (the computational volume for determining the conservation boundaries) was more than 1,000 times larger than the scale of the air behavior that was supposedly being analyzed (Barker *et al.* 1992). The images matched one's expectations of how the air should move, but had little correlation with the actual physics. Tools and codes have become more sophisticated, and most consulting companies have added CFD to their repertoire such that many major building projects, particularly those that involve advanced ventilation schemes, will have CFD analyses performed at some point in the design process. Nevertheless, within the field of building systems, CFD is still treated as a tool for the visualization of coarse air movement, and not as a state-of-the-art method for characterizing the extremely complex physics of discrete air behavior.

The opportunities that other fields and disciplines have developed through the exploitation of CFD have been noticeably absent from architecture. The primary technology for controlling air behavior in buildings has been substantially unchanged for over a century, and design is heavily dependent on strategies based on conceptual understandings that predate current theories of physics. This conceptual understanding was premised on the belief that air behaved as a homogeneous miasma that transported contaminants from outdoors to indoors. Dilution was thus developed as a response to nineteenth-century concerns about the spread of disease in interior environments (Addington 2001, 2003). The HVAC system emerged at the beginning of the twentieth century as the ideal technology for diluting the multiple sources of heat and mass typically found in an interior. No other system was capable of simultaneously mitigating these diverse sources to provide for temperature, humidity, velocity, and air quality control in the quest to provide a homogeneously dilute interior. Both as a technology and a paradigmatic approach, the HVAC system has maintained its hegemony ever since. All other technologies, including those for passive systems, are fundamentally compared against the standard of the dilute interior environment.

While the technology remained static over the course of the last century, the understanding of the physics of air and heat has undergone a radical transformation. Heat transfer and fluid mechanics, the two sciences that govern the behavior of the interior environment, were the last branches of classical physics to develop theoretical structures that could adequately account for generally observable phenomena. The building blocks began with the codification of the Navier–Stokes equations in the mid-nineteenth century, and they fell into place after Ludwig Prandtl first suggested the concept of the boundary layer in 1904. Nevertheless, the solution of the nonlinear partial differential equations wasn't applicable to complex problems until iterative methods began to be employed in the 1950s, leading to the eventual development of CFD in the late 1960s to early 1970s. If the standard definition of technology is that it is the application of physics, then the HVAC system is clearly idiosyncratic in that it predates the understanding of the governing physics. Many might argue that this is irrelevant, regardless of the obsolescence of the technology—it is and will continue to dominate building systems for many years.

The use of CFD has been constrained by the normative understanding of the building technology. This contrasts significantly with its use in other fields in which CFD simulation is *supra* the technology, and not subordinate. For example, in building design, CFD is often used to help determine the optimal location of diffusers and returns for the HVAC system. In engineering, CFD is used to investigate autonomous physical behaviors—such as vortex shedding in relationship to fluid viscosity—and technologies would then be developed to act in accordance with this behavior. Existing building technologies heavily determine the behavior of air, whereas the extant air behavior does not determine the building technology. Many who understand the critical importance of this conceptual distinction would argue, however, that it is of little relevance for architecture. Building technologies have long life spans and do not undergo the rapid cycles of evolution and obsolescence that characterize technological development in the science-driven fields. Given that one cannot radically change building technologies, and, as a result, cannot investigate building air behavior without the overriding influence of HVAC-driven air movement, then how can the discipline of architecture begin to exploit the possibilities inherent in numerical simulation?

6.2 Determining the appropriate problem (what to model)

The methods currently available for characterizing transient fluid behavior—theoretical analysis, empirical evaluation and numerical description—have yet to yield a satisfactory understanding of the air environment in a building. While it is true that room air is representative of a multitude of often-conflicting processes, including forced convection, local buoyancy, radiative effects, and local thermal and moisture generation, these methods have nevertheless been applied with reasonable success for characterizing fluid behavior in many other complex applications. Much of this discrepancy between the success and failure of characterization is likely related to the driving objective behind the studies. Outside of the architecture field, the primary purposes for studying fluid behavior are the identification of the key phenomenological interactions and the determination of the order of magnitude of the significant variables. Even within the complex thermal fields and geometric scales of microelectronics packaging (one of the largest product arenas that utilizes CFD simulation), investigation is directed toward the isolation of a behavior in order to determine the relevant variables for control. Evaluation of room air behavior, however, has typically been concerned with optimizing the selection between standard design practices based on commercially available systems. For example, ASHRAE's project 464, one of the building industry's initial efforts toward codifying CFD procedures, was premised on the assumption that the purpose of CFD is to predict air movement in a room under known conditions (Baker and Kelso 1990). The first major international effort on the validation of CFD methods for room air behavior only considered the following parameters as relevant: HVAC system type, inlet and outlet locations, room proportions and size, furniture and window locations, and the number of computers and people, both smokers and nonsmokers (Chen *et al.* 1992).

Clearly, the foci for CFD simulations of the interior environment are prediction and evaluation—prediction of normative responses, and evaluation of standard systems. Indeed, ASHRAE's current initiative on Indoor Environmental Modeling—Technical Committee 4.10—states that its primary objective is to facilitate the application of simulation methods across the HVAC industry (ASHRAE 1997). These approaches to CFD simulation are quite different from the intentions in the science and engineering realms. A recent keynote address to the applied physics community concluded that the number one recommendation for needed research, development, and implementation issues in computational engineering and physics was that "the application domain for the modeling and simulation capability should be well-understood and carefully defined." (Oberkampf *et al.* 2002). Even though the aerospace discipline pioneered the use of CFD over thirty years ago, and thus has the greatest experience with simulation modeling, aerospace researchers are concerned that it will take another decade just to verify the mathematics in the simulation codes, and even longer to confirm the physical realism of the models (Roache 1998, 2002). NASA describes CFD as an "emerging technology" (NPARC 2003). In stark contrast, a recent review of "The state of the art in ventilation design" concluded that any more attention to the sophisticated algorithms in the CFD world was unnecessary, and that the real need was a software tool that could be picked up quickly (Stribling 2000).

The realm of building simulation considers CFD to be a useful tool for predicting the performance of building systems, particularly HVAC systems. The science and engineering disciplines consider CFD to be a powerful numerical model for studying the behavior of physical processes. These disciplines also recognize an inherent and problematic tautology. Analytical descriptions and empirical evaluations provide the two normative methods of studying physical phenomena. CFD, as the numerical discretization of the governing equations, is thus a subset of the two—based on the analytical description but empirically validated. The tautology emerges when one recognizes that the greatest utility of CFD is for the investigation of problems that can't be empirically tested. As such, many CFD simulations are at best extrapolations—more than sufficient for the investigation of phenomena, insufficient for predicting actual performance.

One could argue that, unlike the science and engineering disciplines in which technologies are contingent on and are developed in response to the identification of new phenomena, the field of building systems has been dominated by a single technological type that has persisted for over a century. This technological type is based on the dilution of heat and mass generation.

The impact of modeling the response of this existing technology, however, brings two problems—the approach to CFD limits the exploration of phenomena, and the privileging of the dilution-based system constrains the modeling type such that anything other than high velocity systems can't easily be examined. By basing the building's performance criteria on the HVAC norm, the resulting simulation models tend toward forced convection—pressure differential is the driving factor—rather than natural convection, or buoyancy, in which density is the driving factor. Yet, buoyant flows predominate in building interiors if one steps back to examine the extant behaviors rather than automatically include the technological response (Table 6.1).

Buoyancy-induced air movement occurs when gravity interacts with a density difference. Within buildings, this density difference is generally caused either by thermal energy diffusion or by moisture diffusion. Surface temperatures—walls, windows, roofs, floors—are almost always different from the ambient temperature such that buoyant flow takes place in the boundary layer along the surfaces. All of the

Table 6.1 The constitutive components of basic buoyant flows

Buoyant flow type	Source geometry	Source type	Architectural examples
Conventional	Vertical surface (infinite)	Isothermal	Interior wall
		Constant flux	Exterior wall
	Vertical surface (finite)	Isothermal	Radiant panel
		Constant flux	Window
	Point (on surface)	Constant flux	Material joint, heat exchanger
Unstable	Horizontal surface (infinite)	Isothermal	Interior floor, ceiling
		Constant flux	Heated floor, ceiling below unheated attic
	Horizontal surface (finite)	Isothermal	Radiant/chilled panels
		Constant flux	Skylights (winter)
	Point (on surface)	Constant flux	Heat exchanger, mounted equipment
	Point (free)	Constant flux	Person, small equipment
Stable	Horizontal surface (finite)	Isothermal	Radiant/chilled panels—reverse orientation
		Constant flux	Skylights (summer)
	Point (free)	Constant flux	Luminaires, heat exchanger

Source: Addington (1997).

building's electrical equipments, including computers, lighting and refrigeration, produce heat in proportion to conversion efficiency, inducing an "unbounded" or free buoyant flow. Processes such as cooking and bathing as well as the metabolic exchanges of human bodies produce both thermal energy and moisture. In general, these types of flows are found near entities surrounded by a thermally stratified environment. The thermal processes taking place in a building that are not density driven result from HVAC systems or wind ventilation, but neither of these are extant—both are technologies or responses intended to mitigate the heat and mass transfer of the density-driven processes.

Buoyancy flows have only begun to be understood within the last 30 years, as they are particularly difficult to investigate in physical models. Much of the research was initiated to study atmospheric processes, which are very large scale flows, and only recently has buoyancy at a small scale—that of a particle—begun to be investigated. Furthermore, buoyant air movement at the small scale was of little interest to the major industries that were employing CFD modeling. It was not until the surge in microelectronics during the last two decades that small-scale buoyancy began to receive the same-attention as compressible hypersonic flows and nuclear cooling. As processor chips became faster, and electronics packages became smaller, heat shedding within and from the package became the limiting factor. Although forced convection with fans had been the standard for package cooling for many years, fans were no longer compatible with the smaller and higher performance electronics. Researchers looked toward the manipulation of buoyant behavior to increase the heat transfer rate, leading to comprehensive investigation of cavity convection, core flows, and the impact of angles on the gravity density interface. As a result, the study of thermal phenomena—in particular, buoyant behavior—rather than the optimization of the technology, has been responsible for revolutionizing the microelectronics industry.

If we increase our focus on the simulation of buoyant behavior, we may begin to be able to characterize the discrete thermal and inertial behavior of interior environments. Rather than undermining the current efforts on performance prediction, this approach would expand the knowledge base of simulation modeling, while opening up new possibilities for technology development. But it requires rethinking the approach toward simulation modeling: the privileging of the HVAC system has affected many aspects of the CFD model—not the least of which is the problem definition. One must also ask fundamental questions about boundaries, scale, and similitude.

6.3 Determining the appropriate modeling criteria (how to model)

The most significant difference between modeling to predict performance and modeling to investigate phenomena is the definition of the problem domain. For many indoor air modelers, the problem domain is coincident with the building's different extents— a room, a group of contiguous rooms, or the building envelope. The geometry of the solid surfaces that enclose air volumes serves as the geometry of the problem. If one accepts that the dilute interior environment is also a well mixed environment, then it is not unreasonable for the domain of the problem to be defined by the building's surfaces. If the environment is not well mixed, and even more so if the overarching forced mixing of the HVAC system is eliminated, then the problem domain must respond to the scale of each behavior, not of the building. The elimination of the dominant mixing behavior should result in an aerodynamically quasi-calm core environment, and therefore each thermal input will behave as an individually bounded phenomenon (Popiolek 1993). Indeed, the growing success of displacement ventilation strategies demonstrates that discrete buoyant behaviors will maintain their autonomy if mixing flows are suppressed. As such, individual phenomena can be explored accurately at length-scales relevant to their operative boundaries. Each behavior operating within a specific environment thus determines the boundary conditions and the length-scale of the characteristic variables.

Boundary conditions are the *sine qua non* of CFD simulation. In fluid flow, a boundary is a region of rapid variation in fluid properties, and in the case of interior environments, the important property is that of density. The greater the variation, the more likely a distinct boundary layer will develop between the two states, and the mitigation of all the state variables—pressure, velocity, density, and temperature— will take place almost entirely within this layer. But a rapid variation in density is problematic in continuum mechanics, and thus boundaries conceptually appear as a discontinuity. In numerical and analytical models, boundary conditions provide the resolution for these discontinuities. In buildings, the common assumption is that solid surfaces are the operative boundaries and thus establish the definitive boundary conditions for the simulation model. Solid surfaces establish but one of the typical boundary conditions present—that of the no-slip condition (the tangential velocity of the fluid adjacent to the surface is equal to the velocity of the surface, which for building surfaces is zero). Much more complicated, and more common, are interface boundaries and far-field boundaries. Interface boundaries occur when adjacent fluids have different bulk properties, and as such, are dynamic and deformable. Far-field boundaries occur when the phenomenon in question is small in relation to the domain extents of the surrounding fluid (Figure 6.1).

In all types of buoyant flow, the temperature and velocity fields are closely coupled. Velocities tend to be quite small such that the momentum and viscous effects are of the same order. As a result, the outer edge along a no-slip surface's boundary layer edge may also be a deformable boundary, particularly if the ambient environment is stratified or unstable. In free buoyant flow, and this is particularly so when there is a quiescent ambient environment, one can consider that far-field boundary conditions prevail. Within the major categories of buoyant flow—conventional, unstable, and stable—any of the three types of boundary conditions may be present. Even for a given flow, small changes in one of the variables may cause the flow to cycle through different phenomena, and thus the same source might have different boundary conditions at different times and at different locations. One of the more remarkable properties of buoyant flows is that the key parameter for determining heat transfer— the characteristic length (L)—is contingent upon the overall type of flow as well as the flow phenomenon at that instant and that location. Just as the boundary conditions are contingent, so too is the characteristic length and thus the heat transfer. This is very different from forced convection, in which the length is a fixed geometric entity. This contingency plays havoc with the simulation of boundary layer transfer, so the majority of CFD simulations for indoor air environments will substitute empirically derived wall functions for the direct determination of the boundary behavior.

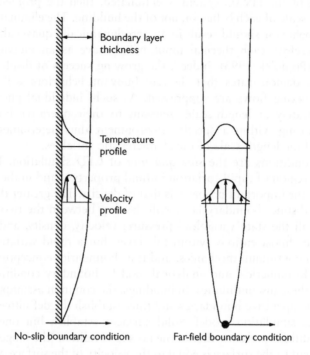

Figure 6.1 Schematic representation of typical buoyant boundary conditions.

In the conventional HVAC system typically used for conditioning interior environments, the length-scales resulting from forced convection are generally an order of magnitude higher than the length-scales from buoyant convection and so buoyant transfer can quite reasonably be approximated from wall functions. As soon as forced convection is removed from the picture, the wall functions currently available are no longer adequate.

Buoyancy forces also directly produce vorticity, and as such, buoyancy-induced flows often straddle the transition point from one flow regime to the other. The precise conditions under which laminar flow becomes unstable has not yet been fully determined, but a reasonable assumption is that buoyancy-induced motion is usually laminar at a length-scale less than one meter, depending on bounding conditions, and turbulent for much larger scale free-boundary flow (Gebhart *et al.* 1988). As neither dominates, and the transition flow is a critical element, then the simulation cannot privilege one or the other. Much discussion is currently taking place within the field of turbulence modeling for interior environments, but the point of controversy is the choice of turbulence models: RANS (Reynolds Averaged Navier–Stokes) in which the $\kappa-\epsilon$ simplification is the most commonly used, or LES (Large Eddy Simulation). Both of these turbulence models are semiempirical. In the $\kappa-\epsilon$ formulation, wall functions must be used. LES numerically models the large turbulent eddies, but treats the small eddies as independent of the geometry at-hand. With fully developed turbulence at high Reynolds numbers, the boundary layers can be neglected and the turbulence can be considered as homogeneous. Again, these turbulent models are adequate for the high Reynolds number flows that forced convection produces, but are entirely inadequate for simulating the laminar to turbulent transition that is chiefly responsible for determining the heat transfer from buoyant flows (Figure 6.2).

In buoyant flow, there is generally no velocity component in the initial conditions. Reynolds similitude by itself cannot adequately describe the flow regime, as turbulence occurs at very low Reynolds numbers. Furthermore, transition between regimes is produced by shifts in the relative strengths of the different forces acting on the flow. The Grashof and Rayleigh numbers, then, are much more meaningful for characterizing buoyant behavior. The Grashof number determines the relative balance between viscous and buoyant forces—essentially the higher the Grashof number, the lower the impact of the solid surfaces (the boundary layers) in restraining buoyant movement.

$$\text{Reynolds number (Re)} = \frac{\rho UL}{\mu} \text{ (ratio of inertial force to viscous force)}$$

$$\text{Grashof number (Gr)} = \frac{g\beta\Delta T L^3}{\nu^2} \text{ (ratio of buoyancy force to viscous force)}$$

A good rule of thumb is that if $Gr/Re^2 \ll 1$, then inertia begins to dominate, and buoyancy can be neglected (Leal 1992). Flows can be considered as homogeneous and the boundary layer can essentially be treated as a wall function. The conventional approach for CFD modeling in building interiors should suffice. If the ratio is O(1), the combined effects of forced and buoyant convection must be considered. And if

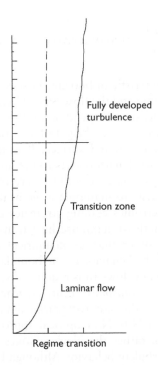

Fully developed
turbulence

Transition zone

Laminar flow

Regime transition

Figure 6.2 Vertical transition from laminar to turbulent regimes in no-slip buoyant flow.

$Gr/Re^2 \gg 1$, then buoyancy dominates. The last condition is the most common for buoyant flows in a quiescent ambient, and as such, boundary layer effects become paramount. Furthermore, diffusion (conduction) from the boundary emerges as an important determinant of the regime of the boundary layer. The Rayleigh number is an indication of the balance between diffusive and buoyant forces: the higher the Rayleigh number, the more likely the boundary layer is turbulent.

$$\text{Rayleigh number (Ra)} = \frac{\rho \beta g \, (T_s - T_f) \, L^3}{\mu \alpha} \text{ (ratio of buoyancy force to diffusion)}$$

Generally accepted ranges of the Rayleigh number for buoyant flow in confined spaces are (Pitts and Sissom 1977):

- conduction regime $\qquad$ $Ra < 10^3$
- asymptotic flow $\qquad$ $10^3 < Ra < 3 \times 10^4$
- laminar boundary layer flow $\quad$ $3 \times 10^4 < Ra < 10^6$
- transition $\qquad$ $10^6 < Ra < 10^7$
- turbulent boundary layer flow $\quad$ $10^7 < Ra$

Given the formulation of the Rayleigh number, it is evident that the typical buoyant flow encountered in buildings will have multiple regimes within its boundary layer, and each regime will have a significantly different heat transfer rate—represented by the Nusselt number.

$$\text{Nusselt number (Nu)} = \frac{hL}{k} \text{ (ratio of convective transport to diffusive transport)}$$

The most interesting characteristic of buoyant flow is that the characteristic length is contingent on both the regime and the flow type, whereas in forced convection, the length is a fixed geometric measure. Depending on the flow description, the characteristic length may be determined by the height of a vertical isothermal surface or the square root of the area of a horizontal surface. If isothermal vertical surfaces are closely spaced, the characteristic length reverts to the horizontal spacing, and particularly interesting relationships emerge if surfaces are tilted (Incropera 1988). As a result, a large opportunity exists to manipulate the heat transfer from any buoyant flow. For example, microelectronics cooling strategies depend heavily on the management of characteristic length to maximize heat transfer to the ambient environment. Clearly, an approach other than semiempirical turbulence models must be found to accurately simulate the behavior of the boundary (Table 6.2).

Accurate modeling of buoyant flows thus requires discrete modeling of the boundary layer. For transition regimes, normally occurring near and above $Ra = 10^7$, Direct Numerical Simulation (DNS) is the only recognized simulation method for determining turbulence (Dubois *et al.* 1999). No empirical approximations or numerical simplifications are used in DNS, rather the Navier–Stokes equations are solved at the length-scale of the smallest turbulent behavior. Although DNS is commonly used in the science community, it has found almost no application in the modeling of room air. The number of nodes needed to discretize all scales of turbulence increases roughly as

Table 6.2 Interdependence of the characteristic length and the flow phenomenon

Buoyant flow type	Flow phenomenon	Variables	Characteristic length
Conventional	Onset of boundary layer flow	Ra	Vertical height of surface L
	Onset of turbulence	x/L, Ra_x	Vertical distance x along surface L
	Core circulation	Stratification (S), Ra_h	Horizontal distance h between vertical surfaces
Unstable	Onset of boundary layer flow	Ra_h	$\sqrt{A}$ (area) of the horizontal surface
	Separation	Ra_h	$\sqrt{A}$ of the horizontal surface
	Onset of turbulence	z/L, Ra_z	Vertical distance z along total height of flow L
	Flow stabilization	Entrainment (E), S, Ra	Total height of flow L
Stable	Onset of boundary layer flow	E, Ra_h	One-half of the shortest horizontal side, or $\sqrt{A}/2$

Source: Addington (1997).

the cube of the Reynolds number. A standard estimate is that if eddies are 0.1–1 mm in size, then the total grid number for a three dimensional air flow is around 10^{11} to 10^{12} and that current supercomputers can only handle a grid resolution of 10^8 (Chen 2001). This widely accepted assumption of the impracticality of DNS has to do with the presumed necessity to model the entire room coupled with the inclusion of high Reynolds number flows. If the entire room or zone with its many surfaces must be included in the model, then CFD modelers must confine the simulation to macro-scale so as to keep the total number of calculation nodes at a reasonable level. The macro-scale requirement not only affects the simulation of turbulence, but it also demands quite large grids for discretization. Ideally, even if a turbulence model is used, the boundary layer should contain 20–30 nodes for a reasonable approximation (Mendenhall *et al.* 2003). The typical building model instead encompasses the entire cross-section of the boundary layer within one much larger node (and this would be a conservative volume, rather than a finite element). Indeed, CFD modelers have been encouraged to simplify models by increasing the scale even further in order to reduce the computational penalty without sacrificing the room geometry.

Does this mean that CFD modelers of interior environments must use DNS if they wish to explore buoyancy and include boundary layer behavior? Not necessarily. But it does mean that some initial groundwork must be laid in the clarification of domains and validation of the individual phenomenon models.

6.4   Determining the appropriate validation strategy

Simulation is a conceptual model of physical reality, and as such comparison with the physical experiment is necessary, or comparison with computational results obtained with mathematical models involving fewer assumptions, such as DNS. These comparisons take the form of validation—ensuring that the right equations are solved—and verification—determining that the equations are solved correctly. Currently, Validation and Verification (V&V) consumes the majority of activities devoted to the development and application of CFD in the science and engineering communities. NASA, as the originator of CFD, stands at the vanguard of the V&V effort, although buoyant and low Reynolds number flows are not a significant part of their focus. The American Institute of Aeronautics and Astronautics (AIAA) has prepared a guide for V&V that is used throughout the engineering communities and has also been adopted by NASA's NPARC Alliance (dedicated to the establishment of a national CFD capability). The American Society of Mechanical Engineers (ASME), as the primary discipline using CFD, has mounted a concerted effort to establish standards and codes for V&V, but it is beginning its efforts on Computational Solid Mechanics (CSM) before turning to CFD.

The application of CFD to interior environments does not fall under the major umbrellas of disciplinary oversight. Furthermore, the burgeoning commercial potential of CFD has led software designers to produce "user-friendly" codes for nonscientists that eliminate many of the difficult steps and decisions in setting up and solving the simulation problem. Meshes can be automatically generated from a geometric model, and defaults exist for the numerical procedures and boundary conditions. Any user with a PC and a commercial CFD code can produce simulations with impressively complex velocity and temperature profiles that may have little requisite relationship to the thermal behavior of the building other than a recognizable geometric section and plan. In response to the flood of CFD codes and consultants inundating the building simulation market, several organizations, including ASHRAE and the International Energy Agency (IEA), followed NASA's lead and launched validation efforts to establish standards and verification for CFD modeling. Their efforts have been thwarted, however, by the overarching assumption that benchmark cases must match the behavior induced by current HVAC technologies in building interiors, thus significantly increasing the specificity of the model. As a result, in spite of the participation of numerous researchers, and the use of several independent yet "identical" test facilities for empirical validation, few applicable conclusions or directions for users have been produced. In 1992, the IEA summarized their work in a database of several hundred precalculated CFD cases in which a single room office with a window had been simulated (Chen *et al.* 1992). Rather than serving as a validation database for comparison, the precalculated cases were intended to supplant CFD modeling by inexperienced users.

Validation is a complex task, as even simple flows are often not correctly predicted by advanced CFD codes (Wesseling 2001). Since simulation is an attempt to model physical reality, then a comparison to that physical reality is a necessity for validation. In many arenas, however, it is tautological: if validation is the matching of simulation results to an empirical test, but CFD is used for problems that can't be empirically evaluated, then which is the independent standard? In addition, the transient nature and complexity of fluid movement is such that even if empirical data is available, it is difficult to tell the difference between empirical error and modeling error. As such, in most sophisticated engineering applications, validation occurs either through similarity analysis or through benchmarking.

Exact solutions are rare in fluid dynamics. Before CFD was available as a method to "solve" the Navier–Stokes equations, many problems were solved by similarity analysis. Similarity analysis is often termed "dimensional analysis" and it is a method for reducing the number of variables required to describe a physical behavior. The fundamental premise is that any mathematical relationship that represents a physical law must be invariant to a transformation of the units. Dimensionless groups, such as the Reynolds number or the Rayleigh number, serve to distinguish types of flow, and thus cannot be tied to any one set of dimensions. With a reduced number of variables, problems can often be reduced to ordinary differential equations, thus dramatically simplifying the solution. The ability of CFD to solve nonlinear partial differential equations may seem to supplant the usefulness of similarity analysis. Nevertheless, a hallmark of a good validation case is its nondimensionality, as it demonstrates that the case adheres fully to physical law.

CFD simulations of interior environments are almost exclusively dimensioned. Not only is each simulation often treated as a unique case in that it represents a specific situation in a particular building, but the combination of multiple behaviors into a single simulation prevents nondimensionalization. Each behavior drives a similarity transformation, such that multiple behaviors will lead to contradictory scaling. As a result, this major method for CFD validation has not been incorporated into the building simulation arsenal.

Benchmarking is the second and somewhat more problematic method for validation. Benchmarks were traditionally physical experiments, although today there is a great deal of argument as to whether empirical error is of the same order as or greater than computational error. The ASME committee studying V&V has concluded that benchmarking is more useful for verification—the determination that the code is being used properly—rather than for validation (Oden 2003). Analytical benchmarks are considered more accurate, but must of necessity be of a single phenomenon in a straightforward and scalable domain.

Both these methods—similarity analysis and benchmarking—require a breaking down of the problem into its smallest and most fundamental behaviors and domains. A valid CFD model could thus be considered as the extents of its validated constituents. The key issue facing CFD modelers trying to examine larger systems is the level at which a model is no longer causally traceable to the discrete behaviors (Roache 1998). Within the field of indoor air modeling, there has not been the longstanding tradition of evaluating single behaviors either through similarity analysis or through discrete physical models, and as a result, CFD modeling operates at the system level without any linkage to a validated basis of fundamentals. Indeed, CFD is used in lieu of other methods rather than being constructed from them. Furthermore, one of the current trends for CFD modeling of interior environments is conflation, which basically expands the simulation even more at the systems level by attempting to tie the results of the CFD model into the boundary conditions for transfer models that determine energy use.

The consequences of disconnecting the CFD model from its fundamental constituents are not so severe. Conventional technologies and normative design are predictable enough and narrow enough that one does not have to do the aggressive validation so necessary for the aerospace and nuclear disciplines. But CFD modeling demands a change if it is to be used for more than this, and particularly if we wish to explore the phenomena and open up the potential for developing new responses and technologies.

6.5 Potential applications

By extending the realm of CFD simulation from the analysis of existing system response to the investigation of extant thermal inputs, several opportunities may emerge, particularly in relationship to micro- versus macro-scale modeling. High momentum air systems (HVAC as well as wind driven) tend to supplant local air movement such that the resulting scale at which the behavior is manifest depends on the length, for example, of the diffuser throw and is therefore relative to room- or macro-scale. As a result, the CFD model must also be macro-scale and thus the thermal boundary layers and the specifics of buoyant flow are not relevant. When the high-momentum system is eliminated from the analysis, individual buoyancy behaviors will predominate, and the discrete boundary becomes significant. The scale of interest for investigating the thermal behavior relates to the thickness of the boundary layer and thus micro-scale. Room-scale is no longer relevant, and the large grid finite volume models typically used to model room air behavior have no application. Buoyancy behavior in buildings has more in common with microelectronic heat transfer than it does with high-momentum air distribution.

One issue that remains regardless as to whether a macro- or micro-model is used is determining the nature and purpose of the ambient surround—the fluid medium. In microelectronics, the operative assumption is that the fluid medium acts as an infinite sink. There is only one objective: the rapid dissipation of heat away from the object. In buildings, however, we have typically assumed that our objective is the maintenance of the fluid medium. The heat transfer from the object is only important insofar as it affects the ambient surround. The thermal objects in a building, however, may include all the physical surfaces—structure, equipment and people—all of which have different thermal production rates and dissipation needs. The inertial mass of the homogeneous fluid medium has been sufficient to absorb these diverse thermal inputs, but it demands that the room or building volume be controlled. While reasonably effective, after all this approach has been used for over a century, it is not only an extremely inefficient means of controlling local heat dissipation from objects, but it also provides at best a compromise—no single object's heat transfer can be optimized. If instead of trying to macro-model the fluid medium, we began to micro-model the boundary layer of the object, we may begin to be able to mitigate or even control the heat transfer from the object without depending on the inertia of the ambient.

Heat transfer is dependent upon characteristic length. Characteristic length is traditionally considered to be a property of an object or a condition of the prevailing flow description. Room walls have specific dimensions and locations; luminaires and computers are built to consistent specifications. Both sources and sinks are relatively fixed and unchanging in the typical building, this then "fixes" the dimensions and the flow patterns, thus predetermining the characteristic length and the resulting heat transfer. Micro-scale modeling, however, allows us to treat the characteristic length as a variable, affording the opportunity to control the heat transfer at the source.

Characteristic length can be readily changed by adjusting the height of an object, even if the total area of the object is maintained. For example, if a window with dimensions of 2 ft high by 1 ft wide were to be rotated 90° such that it became 1 ft high and 2 ft wide, the total heat transfer would be reduced approximately in half. Fourier's Law of heat conduction cannot account for this, nor can the wall functions currently used in macro-scale modeling of no-slip surfaces. It can only be determined from a micro-scale analysis of the window's boundary layer. The heat transfer could further be reduced by an order of magnitude if the height reduction also brought about a regime change from turbulent to laminar flow. In addition, the characteristic length could be altered in the reverse direction to increase the heat transfer from radiant or chilled panels.

More significant, however, and with potentially greater applicability, is the direct manipulation of the characteristic length through an intervention to the flow behavior. Objects and surfaces would not have to be modified or repositioned, rather a careful positioning of an intervention or a response behavior could shift the flow behavior such that a different characteristic length would drive the heat transfer. For example, for a given heated surface, such as a west-facing wall on a summer afternoon, simply shifting the location of the response behavior, whether a cooled surface or a low-momentum air supply, can substantially impact the resulting heat transfer. A chilled ceiling or a cool air supply near the ceiling will result in nearly a 70% greater heat transfer from the heated wall than if the cool sink were to be moved to the floor (Addington 1997). Although a common belief is that cool air when supplied low is least efficient at dissipating heat, it is most efficacious at reducing the amount of heat that must be dissipated to begin with.

The relative location of the cold sink is one of the most straightforward means to manipulate characteristic length and thus heat transfer (Figure 6.3). Although conventional HVAC systems could be modified to allow this shifting, recent developments in micro- and meso-thermal devices may provide the ideal response. Researchers have toyed with these devices for many years, hoping that they would eventually help to replace HVAC systems. In 1994, the Department of Energy stated that its primary research objective was the development of micro- and meso-technologies for heating and cooling (Wegeng and Drost 1994). They had imagined that a micro-heat pump could be assembled in series into a large sheet, much like wallpaper, such that a relatively small surface area of this thin sheet could easily provide enough capacity to heat and cool a building. The initial projections were that $1\,m^2$ of the sheet would be all that was necessary for heating and cooling the typical home. Their expectations for the micro-heat pump capability have been far exceeded, with today's micro-heat pump capable of transferring two orders of magnitude more heat than the basis for their original calculation, yet the project has been stalled.

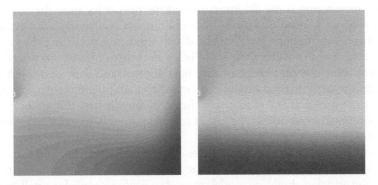

Figure 6.3 Temperature profile comparisons as the relationship between source and sink is shifted. (See Plate V.)

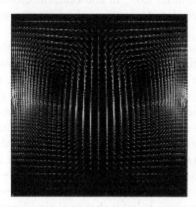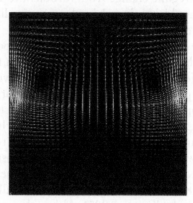

Figure 6.4 Velocity profiles comparing micro-source placement for directly controlling the location of air mixing and/or separation. (See Plate VI.)

The primary concern of the researchers was that the heat pump profile was too small physically to overcome the viscous effects of air in order to provide homogeneous conditions in large volumes. In essence, the hegemony of the HVAC system is such that even radically different technologies are still expected to perform in the same manner.

By characterizing the boundary layer behavior through micro-scale modeling, one could begin to explore the true potential of new technologies. Whereas, the micro- and meso-devices may be impractical as straightforward replacements for the standard components used for conventional HVAC systems, they offer unexploited potential to significantly impact local heat transfer in interior environments. At the scale of the boundary layer, these devices have commensurate length-scales, and as such, are capable of intervening in the layer to effect either a regime change or a shift in the flow phenomenon. The heat transfer rate from any surface or object could then be directly modified without necessitating changes in materials or construction. If the concept is pushed even further, these types of tiny interventions could be effective at forcing particular behaviors in specific locations. For example, a major concern of air quality monitoring is the determination of the proper location for sensors so that they can pick up minute quantities of contaminants. Currently the best method is through increased mixing which has the disadvantage of increasing the contaminant residence time and exposure. One could place the sensor almost anywhere and then, through the judicious placement of tiny heat sources and sinks, establish a specific buoyant plume with thermal qualities designed to manipulate the density of the contaminant to ensure that the sensor sees the contaminant first (Figure 6.4). But these types of solutions, experiments, or just even ideas can only be explored through CFD simulation.

CFD simulation has been a boon to building system designers, and its impact in improving both the efficiency and efficacy of conventional systems cannot be discounted. Nevertheless, building modelers need to begin to consider small-scale behaviors so as to expand the application of CFD from the prediction of the known to the exploration of the unknown.

References

Addington, D.M. (1997). "Boundary layer control of heat transfer in buildings." Harvard University Dissertation.

Addington, D.M. (1998). "Discrete control of interior environments in buildings." In *Proceedings of the 1998 ASME Fluids Engineering Division*. American Society of Mechanical Engineers, New York.

Addington, D.M. (2001). "The history and future of ventilation." *The Indoor Air Quality Handbook*. Mc-Graw-Hill, New York.

Addington, D.M. (2003). "Your breath is your enemy." *Living with the Genie: Governing Technological Change in the 21st Century*. Island Press, Washington, DC.

ASHRAE Technical Committee 4.10 Activities (1997). ASHRAE Insights. September 1997.

Baker, A.J. and Kelso, R.M. (1990). "On validation of computational fluid mechanics procedures for room air motion prediction." *ASHRAE Transactions*, Vol. 96, No. 1, pp. 760–774.

Banks, J. (ed.) (1998). *Handbook of Simulation*. John Wiley & Sons, New York.

Barker, T., Sedgewick, A., and Yau, R. (1992). "From intelligent buildings to intelligent planning." *The Arup Journal*, Vol. 27, No. 3, pp. 16–19.

Beausoleil-Morrison, I. (2001). "Flow responsive Modelling of Internal Surface Convection." In *Seventh International IBPSA Conference*. Rio de Janeiro, Brazil. August 2001.

Chen, Q. (2001). *Indoor Air Quality Handbook*. Mc-Graw-Hill, New York.

Chen, Q., Moser, A., and Suter, P. (1992). *A Database for Assessing Indoor Airflow, Air Quality and Draught Risk*. International Energy Agency, Zurich.

Chung, T.J. (2002). *Computational Fluid Dynamics*. Cambridge University Press, Cambridge, UK.

Dubois, T., Jauberteau, F., and Temam, R. (1999). *Dynamic Multilevel Methods and the Numerical Simulation of Turbulence*. Cambridge University Press, Cambridge, UK.

Elliot, C.D. (1992) *Technics and Architecture*. The MIT Press, Cambridge, MA.

Emmerich, S.J. (1997). *Use of Computational Fluid Dynamics to Analyze Indoor Air Issues*. US Department of Commerce.

Foias, C., Rosa, R., Manley, O., and Temam, R. (2001). *Navier–Stokes Equations and Turbulence*. Cambridge University Press, Cambridge, UK.

Gebhart, B., Jaluria, Y., Mahajan, R.L., and Sammakia, B. (1988). *Buoyancy Induced Flows and Transport*. Hemisphere Publishing Corporation, New York.

Incropera, F.P. (1988). "Convection Heat Transfer in Electronic Equipment cooling." *Journal of Heat Transfer*, Vol. 10, pp. 1097–1111.

Leal, L.G. (1992). *Laminar Flow and Convective Transport Processes*. Butterworth-Heinemann, Boston.

Mendenhall, M.R., Childs, R.E., and Morrison, J.H. (2003). "Best practices for reduction of uncertainty in CFD results." In *41st AIAA Aerospace Sciences Meeting*. Reno, Nevada. January 2003.

NPARC Alliance. (2003). *Overview of CFD Verification and Validation*. http://www.grc.nasa.gov/WWW/wind/valid/tutorial/overview.html

Oberkampf, W.L., Trucano, T.G., and Hirsch, C. (2002). *Foundations for Verification and Validation in the 21st Century Workshop*. October, 2002.

Oden, J.T. (2003). "Benchmarks." American Society of Mechanical Engineers Council on Codes and Standards. http://www.usacm.org/vnvcsm

Pitts, D.R. and Sissom, L.E. (1977). *Outline of Theory and Problems of Heat Transfer*. Mc-Graw-Hill Book Company, New York.

Popiolek, Z. (1993). "Buoyant plume in the process of ventilation—heat and momentum turbulent diffusion." In *Proceedings of Annex-26 Expert Meeting*. Poitiers, France.

Roache, P.J. (1998). *Verification and Validation in Computational Science and Engineering*. Hermosa publishers, Albuquerque, NM.

Roache, P.J. (2002). "Code verification by the method of manufactured solutions." *ASME Journal of Fluids Engineering*, Vol. 124, No. 1, pp. 4–10.

Stribling, D. (2000). *The State of the Art in CFD for Ventilation Design*. Vent, Helsinki.

Tritton, D.J. (1988). *Physical Fluid Dynamics*. Clarendon Press, Oxford, UK.

Wegeng, R.S. and Drost, K. (1994). "Developing new miniature energy systems." *Mechanical Engineering*, Vol. 16, No. 9, pp. 82–85.

Wesseling, P. (2001). *Principles of Computational Fluid Dynamics*. Springer, Berlin.

Chapter 7

Self-organizing models for sentient buildings

Ardeshir Mahdavi

7.1 Introduction

7.1.1 Motivation

Buildings must respond to a growing set of requirements. Specifically, an increasing number of environmental control systems must be made to operate in a manner that is energy-effective, environmentally sustainable, economically feasible, and occupationally desirable. To meet these challenges, efforts are needed to improve and augment traditional methods of building control. This chapter specifically presents one such effort, namely the work on the incorporation of simulation capabilities in the methodological repertoire of building control systems.

7.1.2 Design and operation

The use of performance simulation tools and methods for building design support has a long tradition. The potential of performance simulation for building control support is, however, less explored. We do not mean here the use of simulation for computational evaluation and fine-tuning of building control systems designs. We mean the actual (real-time) support of the building controls using simulation technology (Mahdavi 1997a,b, 2001a; Mahdavi *et al.* 1999a, 2000).

7.1.3 Conventional versus simulation-based control

Conventional control strategies may be broadly said to be "reactive". A thermostat is a classical example: The state of a control device (e.g. a heating system) is changed incrementally in reaction to the measured value of a control parameter (e.g. the room temperature). Simulation-based strategies may be broadly characterized as "proactive". In this case, a change in the state of a control device is decided based on the consideration of a number of candidate control options and the comparative evaluation of the simulated outcomes of these options.

7.1.4 Sentient buildings and self-organizing models

In this contribution we approach the simulation-based building control strategy within the broader concept of "sentient" (self-aware) buildings (Mahdavi 2001b,c;

Mahdavi *et al.* 2001a,b) and self-organizing building models (Mahdavi 2003). We suggest the following working definitions for these terms:

Sentient buildings.   A sentient building is one that possesses an internal representation of its own components, systems, and processes. It can use this representation, among other things, toward the full or partial self-regulatory determination of its own status.

Self-organizing building models.   A self-organizing building model is a complex, dynamic, self-updating, and self-maintaining building representation with instances for building context, structure, components, systems, processes, and occupancy. As such, it can serve as the internal representation of a sentient building toward real-time building operation support (building systems control, facility management, etc.).

Simulation-based building control.   Within the framework of a simulation-based control strategy, control decisions are made based on the comparative evaluation of the simulated implications (predicted future results) of multiple candidate control options.

Note that in this contribution, the terms "sentient" and "self-organizing" are used in a "weak" ("as-if") sense and are not meant to imply ontological identity with certain salient features of biological systems in general and human cognition in particular. Moreover, the core idea of the simulation-based building systems control strategy could be discussed, perhaps, without reference to the concepts of sentient buildings and self-organizing models. However, the realization of the latter concepts is indispensable, if the true potential of simulation technologies for building operation support is to be fully developed.

7.1.5   Overview

Section 7.2 describes the concept of sentient buildings. Section 7.3 is concerned with self-organizing building models. Section 7.4 explains in detail the simulation-based building control strategy and includes descriptions of related prototypical physical and computational implementations. Section 7.5 summarizes the conclusions of the chapter.

7.2   Sentient buildings

A sentient building, as understood in this chapter, involves the following constituents (cp. Figure 7.1):

1   Occupancy—this represents the inhabitants, users, and the visitors of the building.
2   Components, systems—these are the physical constituents of the building as a technical artifact (product).
3   Self-organizing building model—this is the core representation of the building's components, systems, and processes. It provides a sensor-supported continuously updated depiction of the actual status of the occupancy and the building (with its components and systems), as well as the immediate context (environment) in

which the building is situated. To be operationally effective, it is updated fairly autonomously based on pervasive sensor-supported data collection and algorithms for the interpretation of such data.

4   Model-based executive unit—this constitutes the evaluative and decision-making agency of the sentient building. Simulation-based control strategies are part of this unit's repertoire of tools and methods for decision-making support.

Depending on the specific configuration and the level of sophistication of a sentient building, occupants may directly manipulate the control devices or they may request from the executive unit the desired changes in the state of the controlled entity. Likewise, the executive unit may directly manipulate control devices or suggest control device manipulations to the users. As such, the division of the control responsibility between the occupants and the executive unit can be organized in very different ways. Nonetheless, some general principles may apply. For instance, it seems appropriate that the occupants should have control over the environmental conditions in their immediate surroundings. Moreover, they should be given certain override possibilities, in case the decisions of the automated building control systems should disregard or otherwise interfere with their preferred indoor environmental conditions. On the other hand, the executive unit needs to ensure the operational integrity and efficiency of the environmental systems of the building as a whole. It could also fulfill a negotiating role in cases where user requirements (e.g. desired set-points for indoor environmental parameter) would be in conflict with each other.

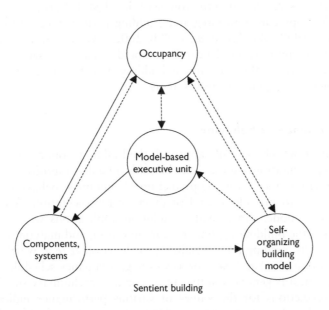

Figure 7.1 Scheme of the constitutive ingredients of a sentient building (continuous lines: control actions; dashed lines: information flows).

7.3   Self-organizing models

7.3.1   Requirements

To serve effectively as the representational core of a sentient building, a self-organizing model must fulfill at least two requirements. First, such a model must incorporate and integrate both a rather static building product view and a rather dynamic behavioral view of the building and its environmental systems. Second, to provide real-time building operation support, the model must be easily adaptable, that is, it must respond to changes in occupancy, systems, and context of the building. Ideally, the model should detect and reflect such changes automatically, that is, it must update (organize) itself autonomously (without intervention by human agents).

7.3.2   Building as product

Numerous representational schemes (product models) have been proposed to describe building elements, components, systems, and structures in a general and standardized manner. Thereby, one of the main motivations has been to facilitate hi-fidelity information exchange between agents involved in the building delivery process (architects, engineers, construction people, manufacturers, facility managers, users). A universal all-purpose product model for buildings has not emerged and issues such as model integration across multiple disciplines and multiple levels of informational resolution remain unresolved (Mahdavi 2003). Nonetheless, past research has demonstrated that integrated building representations may be developed, which could support preliminary simulation-based building performance evaluation. An instance of such a representation or a shared building model (see Figure 7.2) was developed in the course of the SEMPER project, a research effort toward the development of an integrated building performance simulation environment (Mahdavi 1999; Mahdavi et al. 1999b, 2002). We submit here, without proof, that such a shared building model can be adapted as part of a self-organizing building model and provide, thus, a sentient building with the requisite descriptions of building elements, components, and systems.

7.3.3   Performance as behavior

Building product models typically approach the building from a "timeless" point of view. Their representational stance may be said to be decompositional and static. In contrast, simulation allows for the prediction of buildings' behavior over time and may be thus said to provide a kind of dynamic representation. A comprehensive building product model can provide simulation applications with necessary input data concerning the building's geometry, configuration, and materials. This information, together with assumptions pertaining to the context (e.g. weather conditions, available solar radiation) and basic processes (e.g. occupancy schedules, lighting, and ventilation regimes) is generally sufficient to conduct preliminary simulation studies resulting in predictions for the values of various performance indicators such as energy use and indoor temperatures. However, more sophisticated simulations involving detailed behavior of a building's environmental systems and their control

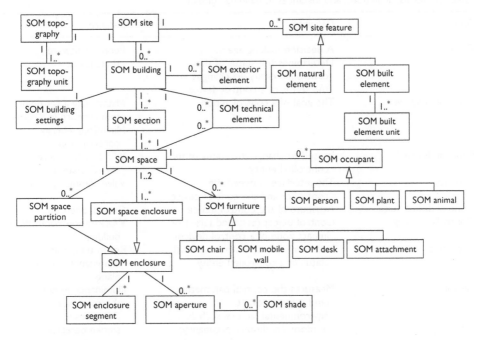

Figure 7.2 SEMPER's shared object model (SOM).

processes (heating, cooling, ventilation, and lighting) require a richer kind of underlying representational framework. Such representation must combine detailed building product information with building control process modeling.

7.3.4   Control as process

There appears to be a divide between modes and styles of control system representation in the building control industry and representational habits in architecture and building science. Specifically, there is a lack of systematic building representations that would unify product model information, behavioral model information, and control process model information. To better illustrate this problem and possible remedies, first some working definitions regarding the building control domain are suggested (see Table 7.1). These definitions are neither definitive nor exclusive, but they can facilitate the following discussions.

A basic control process involves a controller, a control device, and a controlled entity (see Figure 7.3). An example of such a process is when the occupant (the controller) of a room opens a window (control device) to change the temperature (control parameter) in a room (controlled entity). Note that such process may be structured recursively, so that an entity that might be seen as device at a "higher" level may be seen as a controlled entity at a "lower level". For example, when a control algorithm (controller) instructs a pneumatic arm to close (i.e. change the state of) a window, one could presumably argue that the pneumatic arm is a control device

Table 7.1 Terms, definitions, and instances in building control

Term	Definition	Instance
Controller	A decision-making agent. Determines the status of the controlled entity via changes in the status of a control device	People, software, thermostat
Control objective	The goal of a control action	Maintaining a set-point temperature in a room Minimizing energy consumption
Control device	Is used to change the status of the controlled entity	Window, luminaire, HVAC system
Actuator	The interface between the controller and the control device	Valve, dimmer, people
Control device state	Attribute of the control device	Closed, open, etc.
Controlled entity	Control object (assumed target or impact zone of a control device)	Workstation, room, floor, building
Control parameter	Indicator of the (control-relevant) status of a controlled entity	Room temperature, illuminance on a working plane
Sensor	Measures the control parameter (and other relevant environmental factors, such as outdoor conditions, occupancy); reports the status of a control device	Illuminance meter, thermometer, CO_2-sensor, smoke detector, electricity counter
Control action	Instructed change in the status of a control device targeted at changing the status of a controlled entitiy	Opening of a window, changing the status of a dimmer
Control state space	The logical space of all possible states (positions) of a (or a number of) control device(s)	The temperature range of a thermostat, the opening range of a valve

Source: Mahdavi (2001b,c).

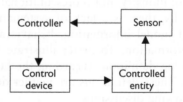

Figure 7.3 A general control scheme.

and the window the controlled entity. To avoid confusion, though, we prefer here to reserve the term controlled entity for the "ends" of the control process. Since opening and closing a window is not an end on itself but a means to another end (e.g. lowering the temperature of a space), we refer to the window as a device and not as the controlled entity.

As we shall see, the basic control process model depicted in Figure 7.3 is highly schematic and must be substantially augmented, as soon as realistic control processes are to be represented. Nonetheless, it makes sense at this point to explore ways of coupling this basic process model with an instance of the previously mentioned building product models. If properly conceived, such a unified building product and process model could act as well as the representational core of a sentient building.

Figure 7.4 illustrates a high-level expression of such a combined building product and control model. While certain instances of the product model such as building, section, space, and enclosure constitute the set of controlled entities in the process view, other instances such as aperture or technical systems and devices fulfill the role of control devices.

7.3.5   Control system hierarchy

As mentioned earlier, the primary process scheme presented in Figure 7.3 is rather basic. Strictly speaking, the notion of a "controller" applies here only to a "device controller" (DC), that is, the dedicated controller of a specific device. The scheme stipulates that a DC receive control entity's state information directly from a sensor, and, utilizing a decision-making functionality (e.g. a rule or an algorithm that encapsulates the relationship between the device state and its sensory implication), sets the state of the device. Real-world building control problems are, however, much more complex, as they involve the operation of multiple devices for each environmental system domain and multiple environmental system domains (e.g. lighting, heating, cooling, ventilation).

As such, the complexity of building systems control could be substantially reduced, if distinct processes could be assigned to distinct (and autonomous) control loops. In

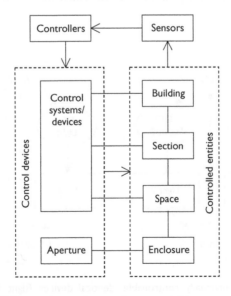

Figure 7.4 A high-level building product and control process scheme.

practice, however, controllers for various systems and components are often interdependent. A controller may need the information from another controller in order to devise and execute control decisions. For example, the building lighting system may need information on the building's thermal status (e.g. heating versus cooling mode) in order to identify the most desirable combination of natural and electrical lighting options. Moreover, two different controllers may affect the same control parameter of the same impact zone. For example, the operation of the window and the operation of the heating system can both affect the temperature in a room. In such cases, controllers of individual systems cannot identify the preferable course of action independently. Instead, they must rely on a higher-level controller instance (a "meta-controller", (MC) as it were), which can process information from both systems toward a properly integrated control response.

We conclude that the multitude of controllers in a complex building controls scheme must be coupled appropriately to facilitate an efficient and user-responsive building operation regime. Thus, control system features are required to integrate and coordinate the operation of multiple devices and their controllers. Toward this end, control functionalities must be distributed among multiple higher-level controllers or MCs in a structured and distributed fashion. The nodes in the network of DCs and MCs represent points of information processing and decision-making.

In general, "first-order" MCs are required: (i) to coordinate the operation of identical, separately controllable devices and (ii) to enable cooperation between different devices in the same environmental service domain. A simple example of the first case is shown in Figure 7.5 (left), where an MC is needed to coordinate the operation of two electric lights to achieve interior illuminance goals in a single control zone. In the second case (see Figure 7.5, right), movable blinds and electric lights are coordinated to integrate daylighting with electric lighting.

In actual building control scenarios, one encounters many different combinations of the cases discussed here. Thus, the manner in which the control system functionality

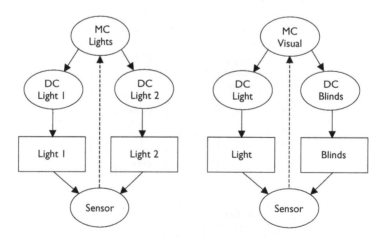

Figure 7.5 Left: MC for individually controllable identical devices; Right: MC for different devices addressing the same control parameter.

is distributed among the controllers must be explicitly organized. The control process model must be created using a logical, coherent, and reproducible method, so that it can be used for a diverse set of building control applications. Ideally, the procedure for the generation of such a control process model should be automated, given its complexity, and given the required flexibility, to dynamically accommodate changes over time in the configuration of the controlled entities, control devices, and their respective controllers.

7.3.6 Automated generation of control system representation

We have developed and tested a set of constitutive rules that allow for the automated generation of the control system model (Mahdavi 2001a,b). Such a model can provide a template (or framework) of distributed nodes which can contain various methods and algorithms for control decision-making. Specifically, five model-generation rules are applied successively to the control problem, resulting in a unique configuration of nodes that constitute the representational framework for a given control context. The first three rules are generative in nature, whereas rules 4 and 5 are meant to ensure the integrity of the generated model. The rules may be stated as follows:

1 Multiple devices of the same type that are differentially controllable and that affect the same sensor necessitate an MC.
2 More than one device of different types that affect the same sensor necessitates an MC.
3 More than one first-order MC affecting the same device controller necessitates a second-order (higher-level) MC.
4 If in the process a new node has been generated whose functionality duplicates that of an existing node, then it must be removed.
5 If rule 4 has been applied, any resulting isolated nodes must be reconnected.

The following example illustrates the application of these rules (Mertz and Mahdavi 2003). The scenario includes two adjacent rooms (see Figure 7.6), each with four luminaires and one local heating valve, which share an exterior movable louvers. Hot water is provided by the central system, which modulates the pump and valve state to achieve the desired water supply temperature. In each space, illuminance and temperature is to be maintained within the set-point range. This configuration of spaces and devices stems from an actual building, namely the Intelligent Workplace (IW) at Carnegie Mellon University, Pittsburgh, USA (Mahdavi et al. 1999c).

One way of approaching the definition of control zones (controlled entities) is to describe the relationship between the sensors and devices. From the control system point of view, controlled entities are "represented" by sensors, and the influence of devices on the controlled entities is monitored via sensory information. In the present example, an interior illuminance sensor (E) and a temperature sensor (t) are located in each space. The sensors for Space-1 are called E_1 and t_1, and those for Space-2 are called E_2 and t_2. In Space-1, both the louvers and electric lights can be used to meet the illumination requirements. As shown in Figure 7.7, sensor E_1 is influenced by the louver state, controlled by DC-Lo1, as well as by the state of four electric lights, each

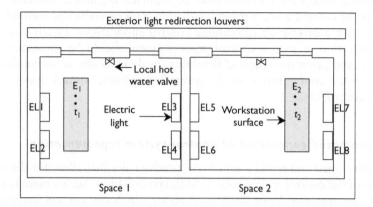

Figure 7.6 Schematic floor plan of the test spaces.

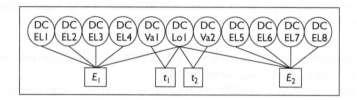

Figure 7.7 Association between sensors and devices (cp. text and Figure 7.6).

controlled by a DC-EL. Similarly, both the local valve state and the louver state influence the temperature in Space-1 (t_1). Analogous assumptions apply to Space-2.

Once the control zones (controlled entities) have been defined, the generation rules can be applied to the control problem as illustrated in Figure 7.7, resulting in the representation of Figure 7.8. A summary of the application of rules 1, 2, and 3 in this case is shown in Table 7.2. As to the application of rule 1, four nodes, namely DC-EL1, EL2, EL3, and EL4 are of the same device type and all impact sensor E_1. Thus, an MC is needed to coordinate their action: MC-EL_1. Similarly, regarding the application of rule 2, both DC-Lo1 and DC-Va1 impact the temperature of Space-1. Thus, MC-Lo_Va_1 is needed to coordinate their action. As to rule 3, four MC nodes control the DC-Lo1 node. Thus, their actions must be coordinated by an MC of second order, namely MC-II EL_Lo_Va_1.

In the above example, rules 1, 2, and 3 were applied to the control problem to construct the representation. Using this methodology, a scheme of distributed, hierarchical control nodes can be constructed. In certain cases, however, the control problem contains characteristics that cause the model not to converge toward a single top-level controller. In these cases, rules 4 and 5 can be applied to ensure convergence. Rule 4 is used to ensure that model functionality is not duplicated. Thereby, the means of detecting a duplicated node lies in the node name. Since the application of rule 4 may

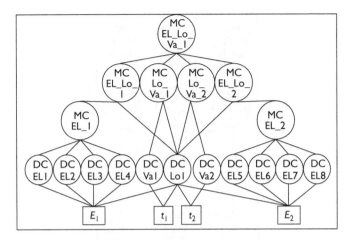

Figure 7.8 An automatically generated control model (cp. text and Figures 7.6 and 7.7).

Table 7.2 Application of rules 1, 2, and 3 (cp. text and Figure 7.8)

Multiple controllers	Affected sensor	Affected device	Meta-controller
Application of rule 1			
EL1, EL2, EL3, EL4	E_1	N/A	MC-EL_1
EL5, EL6, EL7, EL8	E_2	N/A	MC-EL_2
Application of rule 2			
Lo1, VA1	t_1	N/A	MC-Lo_VA_1
Lo1, VA2	t_2	N/A	MC-Lo_VA_2
EL_1, Lo1	E_1	N/A	MC-EL_Lo_1
EL_2, Lo1	E_2	N/A	MC-EL_Lo_2
Application of rule 3			
EL_Lo_1, EL_Lo_2,	N/A	Lo1	MC-II
Lo_VA_1, Lo_VA_2			EL_Lo_VA_1

create hierarchically isolated nodes, rule 5 is applied to reconnect such nodes. The following example illustrates the application of these two rules.

Figure 7.9 shows a model that was constructed using rules 2 and 3. The application of these rules is summarized in Table 7.3. Rule 1 does not apply in this case because there are three distinct device types involved. As to the application of rule 2, DC-EL1 and DC-BL1 both impact the state of E_1 and thus MC-BL_EL_1 is needed to negotiate between them. Three MC nodes are created in this manner. When rule 3 is applied, three second-order MCs are created. It is apparent that the model will not converge. Moreover, the three nodes have the same name: MC-BL_EL_Lo. This is an indication of duplicated functionality (of coordinating devices BL, EL, and Lo). Thus, applying rule 4, nodes MC-BL_EL_Lo_2 and MC-BL_EL_Lo_3 are removed, and applying rule 5, node MC-BL_Lo_1, which is left without a parent node, is connected to the MC-BL_EL_Lo_1.

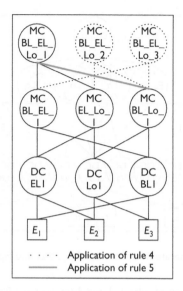

Figure 7.9 Application of rules 4 and 5 (cp. text).

Table 7.3 Application of rules 2 and 3 (cp. text and Figure 7.9)

Multiple controllers	Affected sensor	Affected device	Meta-controller
Application of rule 2			
EL1, BL1	E_1	N/A	MC-BL_EL_1
EL1, Lo1	E_2	N/A	MC-EL_Lo_1
BL1, Lo1	E_3	N/A	MC-BL_Lo_1
Application of rule 3			
BL_EL_1, EL_Lo_1	E_1	DC-EL1	MC-BL_EL_Lo_1
EL_Lo_1, BL_Lo_1	E_2	DC-Lo1	MC-BL_EL_Lo_2
BL_EL_1, BL_Lo_1	E_3	DC-BL1	MC-BL_EL_Lo_3

7.3.7   Real-time model updating

Once a building model is available with instances for building context, structure, systems, status, processes, and occupancy, it can be used to support the real-time building operation (building systems control, facility management, etc.). However, given the complexity of such a model, it seems clear that it needs to be self-organizing, that is, it must maintain and update itself fairly autonomously. Depending on the type and the nature of the entity, system, or process to be monitored, various sensing technologies can be applied to continuously update the status of a building model:

1   Information about critical attributes of external microclimate (e.g. outdoor air temperature, relative humidity, wind speed and direction, global and diffuse irradiance and illuminance) can be gained via a number of already existing sensor

technologies (Wouters 1998; Mahdavi *et al.* 1999c). A compact and well-equipped weather station is to be regarded as a requisite for every sentient building.

2   The success of indoor environmental control strategies can be measured only when actual values of target performance variables are monitored and evaluated. Also in this case there exists a multitude of sensor-based technologies to capture factors such as indoor air temperature, mean radiant temperature, relative humidity, air movement, CO_2 concentration, and illuminance. Further advances in this area are desirable, particularly in view of more cost-effective solutions for embodied high-resolution data monitoring and processing infrastructures.

3   Knowledge of the presence and activities of building occupants is important for the proper functionality of building operation systems. Motion detection technologies (based on ultrasound or infrared sensing) as well as machine vision (generation of explicit geometric and semantic models of an environment based on image sequences) provide possibilities for continuous occupancy monitoring.

4   The status of moveable building control components (windows, doors, openings, shading devices, etc.) and systems (e.g. actuators of the building's environmental systems for heating, cooling, ventilation, and lighting) can be monitored based on different techniques (contact sensing, position sensing, machine vision) and used to update the central building model.

5   Certain semantic properties (such as light reflection or transmission) of building elements can change over time. Such changes may be dynamically monitored and reflected in the building model via appropriate (e.g. optical) sensors.

6   Changes in the location and orientation of building components such as partitions and furniture (due, e.g. to building renovation or layout reconfiguration) may be monitored via component sensors that could rely on wireless ultrasound location detection, utilize radio frequency identification (RFID) technology (Finkenzeller 2002), or apply image processing (De Ipina *et al.* 2002). Gaps in the scanning resolution and placement of such sensors (or cameras) could be compensated, in part, based on geometric reasoning approaches (possibly enhanced through artificial intelligence methods). Moreover, methods and routines for the recognition of the geometric (and semantic) features of complex built environments can be applied toward automated generation and continuous updating of as-is building models (Eggert *et al.* 1998; Faugeras *et al.* 1998; Broz *et al.* 1999).

7.4   A simulation-based control strategy

7.4.1   Introductory remark

We argued that the nodes in the network of DCs and MCs in a building's control scheme represent points of information processing and decision-making. An important challenge for any building control methodology is to find effective methods of knowledge encapsulation and decision-making in such nodes. There are various ways of doing this (Mahdavi 2001a). The simulation-based control method is discussed in the following section. This method can be effectively applied, once the main requirement for the realization of a sentient building is met, namely the presence of a unified building product and process model that can update itself dynamically and autonomously.

7.4.2   Approach

Modern buildings allow, in principle, for multiple ways to achieve desired environmental conditions. For example, to provide a certain illuminance level in an office, daylight, electrical light, or a combination thereof can be used. The choice of the system(s) and the associated control strategies represent a nontrivial problem since there is no deterministic procedure for deriving a necessary (unique) state of the building's control systems from a given set of objective functions (e.g. desirable environmental conditions for the inhabitants, energy and cost-effectiveness of the operation, minimization of environmental impact).

Simulation-based control can potentially provide a remedy for this problem (Mahdavi 1997a, 2001a; Mahdavi et al. 1999a, 2000). Instead of a direct mapping attempt from the desirable value of an objective function to a control systems state, the simulation-based control adopts an "if-then" query approach. In order to realize a simulation-based building systems control strategy, the building must be supplemented with a multi-aspect virtual model that runs parallel to the building's actual operation. While the real building can only react to the actual contextual conditions (e.g. local weather conditions, sky luminance distribution patterns), occupancy interventions, and building control operations, the simulation-based virtual model allows for additional operations: (a) the virtual model can move backward in time so as to analyze the building's past behavior and/or to calibrate the program toward improved predictive potency; (b) the virtual model can move forward in time so as to predict the building's response to alternative control scenarios. Thus, alternative control schemes may be evaluated, and ranked according to appropriate objective functions pertaining to indoor climate, occupancy comfort, as well as environmental and economic considerations.

7.4.3   Process

To illustrate the simulation-based control process in simple terms, we shall consider four process steps (cp. Table 7.4):

1   The first step identifies the building's control state at time t_i within the applicable control state space (i.e. the space of all theoretically possible control states). For clarity of illustration, Table 7.4 shows the control state space as a three-dimensional space. However, the control state space has as many dimensions as there are distinct controllable devices in a building.
2   The second step identifies the region of the control state space to be explored in terms of possible alternative control states at time t_{i+1}.
3   The third step involves the simulation-based prediction and comparative ranking of the values of pertinent performance indicators for the corpus of alternative identified in the second step.
4   The fourth step involves the execution of the control action, resulting in the transition of control state of the building to a new position at time t_{i+1}.

7.4.4   An illustrative example

Let us go through the steps introduced in Section 7.4.3 using a demonstrative experiment regarding daylighting control in an office space in the previously mentioned IW

Table 7.4 Schematic illustration of the simulation-based control process

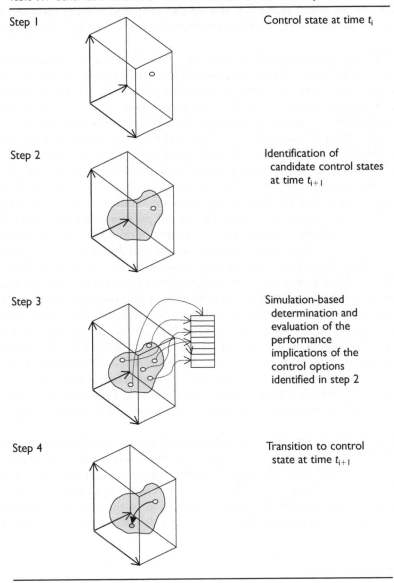

Step I	Control state at time t_i
Step 2	Identification of candidate control states at time t_{i+1}
Step 3	Simulation-based determination and evaluation of the performance implications of the control options identified in step 2
Step 4	Transition to control state at time t_{i+1}

(see Figure 7.6). About 60% of the external wall of this space consists of glazing. The facade system includes a set of three parallel external moveable louvers, which can be used for shading and—to a certain degree—for light redirection. These motorized louvers can be rotated anti-clockwise from a vertical position up to an angle of 105°. We installed an array of 12 illuminance sensors in the central axis of this space at a height of about 0.8 m above the floor to monitor the spatial distribution of the interior illuminance. Outdoor light conditions were monitored using 11 illuminance and irradiance sensors that were installed on the daylight monitoring station on the roof of the IW. As an initial feasibility test of the proposed simulation-based control approach, we considered the problem of determining the "optimal" louver position.

Step 1. In this simple case, the control state space has just one dimension, that is, the position of the louver. We further reduced the size of this space, by allowing only four discrete louver positions, namely 0° (vertical), 30°, 60°, and 90° (horizontal).

Step 2. Given the small size of the control state space in this case, we considered all four possible louver positions as potential candidates to be compared.

Step 3. LUMINA (Pal and Mahdavi 1999), the lighting simulation application in SEMPER (Mahdavi 1999), was used for the prediction of light levels in the test space. LUMINA utilizes the three-component procedure (i.e. the direct, the externally reflected, and the internally reflected component), to obtain the resultant illuminance distribution in buildings. The direct component is computed by numerical integration of the contributions from all of those discretized patches of the sky dome that are "visible" as viewed from reference receiver points in the space. Either computed or measured irradiance values (both global horizontal and diffuse horizontal irradiance) can be used to generate the sky luminance distribution according to the Perez model (Perez *et al.* 1993). External obstruction (i.e. light redirection louvers) are treated by the projection of their outline from each reference point on to the sky dome and the replacement of the relative luminance values of the occupied sky patches with those of the obstruction. A radiosity-based approach is adopted for computing the internally reflected component. The results generated by LUMINA have shown to compare favorably with measurements in several rooms (Pal and Mahdavi 1999). In the present case, measured irradiance values were used at every time-step to generate the sky model in LUMINA for the subsequent time-step. However, trend-forecasting algorithms could be used to predict outdoor conditions for future time-steps.

For each time-step the simulation results (mean illuminance and uniformity levels on a horizontal plane approximately 1 m above the floor) were ordered in a table, which was used to rank and select the most desirable control scenario based on the applicable objective functions. Two illustrative objective functions were considered. The first function aims at minimizing the deviation of the average (daylight-based) illuminance level E_m in the test space from a user-defined target illuminance level E_t (say 500 lx):

$$\text{Minimize } (|E_t - E_m|) \tag{7.1}$$

The second objective function aims at maximizing the uniformity of the illuminance distribution in the test space as per the following definition (Mahdavi and Pal 1999):

$$\text{Maximize } U, \text{ where } U = E_m \cdot (E_m + E_{sd})^{-1} \tag{7.2}$$

Here E_m and E_{sd} are the mean and standard deviation of the illuminance levels measured at various locations in the test space.

At time interval t_i, the simulation tool predicted for four candidate louver positions the expected interior illuminance levels for the time interval t_{i+1} (test space geometry and photometric properties, as well as the outdoor measurements at time interval t_i were used as model input). Based on the simulation results and objective functions, it was possible to determine for each time-step the louver position that was considered most likely to maximize the light distribution uniformity or to minimize the deviation of average illuminance from the target value.

Step 4. Device controller instructed the control device (louver) to assume the position identified in step 3 as most desirable.

To evaluate the performance of the simulation-based control approach in this particular case, we measured during the test period at each time-step the resulting illuminance levels sequentially for all four louver positions and for all selected time intervals. To numerically evaluate the performance of this simulation-based control approach via a "control quality index", we ranked the resulting (measured) average illuminance and the uniformity according to the degree to which they fulfilled the objective functions. We assigned 100 points to the instances when the model-based recommendation matched the position empirically found to be the best. In those cases where the recommendation was furthest from the optimal position, control quality index was assumed to be zero. Intermediate cases were evaluated based on interpolation. Control quality index was found to be 74 for illuminance and 99 for uniformity. The better performance in the case of the uniformity indicator is due to the "relative" nature of this indicator, which, in contrast to the illuminance, is less affected by the absolute errors in the predictions of the simulation model.

7.4.5   Challenges

7.4.5.1   Introduction

In previous sections we described the simulation-based strategy toward building systems control and how this approach, supported by a self-organizing building model, could facilitate the operation of a sentient building. The practical realization of these methods and concepts, however, requires efficient solutions for various critical implementation issues. The appendices of the chapter include case studies involving demonstrative implementation efforts that illustrate some of these problems and their potential solutions.

There are two basic problems of the proposed approach, which we briefly mention but will not pursue in detail, as they are not specific to simulation-based control methodology but represent basic problems related to simulation methods and technologies in general:

1   First, the reliability of simulation algorithms and tools is always subject to validation, and this has been shown to be a difficult problem in the building performance simulation domain. In the context of sentient building implementations, there is an interesting possibility to improve on the predictive capability of the simulation applications by "on-line" calibration of simulation results. This can be done by continuous real-time monitoring of the performance indicator values (using a limited number of strategically located sensors) and comparing those with corresponding simulation results. Using the results of this comparison,

appropriate correction factors may be derived based on statistical methods and neural network applications.

2   Second, preparation of complete and valid input data (geometry, materials, system specifications) for simulation is often a time-consuming and error-prone task. In the context of self-organizing models, however, such data would be prepared mostly in an automated (sensor-based) fashion, thus reducing the need for human intervention toward periodic updating of simulation models.

In the following discussion, we focus on arguably the most daunting problem of the simulation-based control strategy, namely the rapid growth of the size of the control state space in all those cases where a realistic number of control devices with multiple possible positions are to be considered.

Consider a space with n devices that can assume states from s_1 to s_n. The total number, z, of combinations of these states (i.e. the number of necessary simulation runs at each time-step for an exhaustive modeling of the entire control state space) is thus given by:

$$z = s_1, s_2, ..., s_n \tag{7.3}$$

This number represents a computationally insurmountable problem, even for a modest systems control scenario involving a few spaces and devices: An exhaustive simulation-based evaluation of all possible control states at any given time-step is simply beyond the computational capacity of currently available systems. To address this problem, multiple possibilities must be explored, whereby two general approaches may be postulated, involving: (i) the reduction of the size of the control state space region to be explored, (ii) the acceleration of the computational assessment of alternative control options.

7.4.5.2   The control state space

At a fundamental level, a building's control state space has as many dimensions as there are controllable devices. On every dimension, there are as many points as there are possible states of the respective device. This does not imply, however, that at every time-step the entire control state space must be subjected to predictive simulations.

The null control state space.   Theoretically, at certain time-steps, the size of the applicable control state space could be reduced to zero. Continuous time-step performance modeling is not always necessary. As long as the relevant boundary conditions of systems' operation have remained either unchanged or have changed only insignificantly, the building may remain in its previous state. Boundary conditions denote in this case factors such as outdoor air temperature, outdoor global horizontal irradiance, user request for change in an environmental condition, dynamic change in the utility charge price for electricity, etc. Periods of building operation without significant changes in such factors could reduce the need for simulation and the associated computational load.

Limiting the control state space.   Prior to exhaustive simulation of the theoretically possible control options, rules may be applied to reduce the size of the control state space to one of practical relevance. Such rules may be based on heuristic and logical reasoning. A trivial example of rules that would reduce the size of the control state space would be to exclude daylight control options (and the corresponding simulation runs) during the night-time operation of buildings' energy systems.

Compartmentalization. The control state space may be structured hierarchically, as seen in Section 7.3. This implies a distribution of control decision-making across a large number of hierarchically organized decision-making nodes. We can imagine an upward passing of control state alternatives starting from low-level DCs to upper-level MCs. At every level, a control node accesses the control alternatives beneath and submits a ranked set of recommendations above. For this purpose, different methods may be implemented in each node, involving rules, tables, simulations, etc. Simulation routines thus implemented, need not cover the whole building and all the systems. Rather, they need to reflect behavioral implications of only those decisions that can be made at the level of the respective node.

"Greedy" navigation and random jumps. Efficient navigation strategies can help reduce the number of necessary parametric simulations at each time-step. This is independent of the scale at which parametric simulations are performed (e.g. whole-building simulation versus local simulations). In order to illustrate this point, consider the following simple example: Let D be the number of devices in a building and P the number of states each device can assume. The total number z of resulting possible combinations (control states) is then given by Equation (7.4).

$$z = P^D \tag{7.4}$$

For example, for $D = 10$ and $P = 10$, a total of 10 billion possible control states results. Obviously, performing this number of simulations within a time-step is not possible. To reduce the size of the segment of the control state space to be explored, one could consider, at each time-step, only three control states for each device, namely the status quo, the immediate "higher" state, and the immediate "lower" state. In our example, this would mean that $D = 10$ and $P = 3$, resulting in 59,049 control states. While this result represents a sizable reduction of the number of simulation, it is still too high to be of any practical relevance. Thus, to further reduce the number of simulations, we assume the building to be at control state A at time t_1. To identify the control state B at time t_2, we scan the immediate region of the control state space around control state A. This we do by moving incrementally "up" and "down" along each dimension, while keeping the other coordinates constant. Obviously, the resulting number of simulations in this case is given by:

$$z = 2D + 1 \tag{7.5}$$

In our example, $D = 10$. Thus, $n = 21$. Needless to say, this number represents a significantly more manageable computational load. However, this "greedy" approach to control state space exploration obviously bears the risk that the system could be caught in a performance state corresponding to a local minima (or maxima). To reduce this risk, stochastically based excursions to the more remote regions of the control state space can be undertaken. Such stochastic explorations could ostensibly increase the possibility of avoiding local minima and maxima in search for optimal control options.

7.4.5.3   Efficient assessment of alternative control options

Our discussions have so far centered on the role of detailed performance simulation as the main instrument to predict the behavior of a building as the result of alternative control actions. The obvious advantage of simulation is that it offers the possibility of an explicit analysis of various forces that determine the behavior of the building. This explicit modeling capability is particularly important in all those cases, where multiple environmental systems are simultaneously in operation. The obvious downside is that detailed simulation is computationally expensive. We now briefly discuss some of the possible remedies.

Customized local simulation.   As mentioned earlier, simulation functionality may be distributed across multiple control nodes in the building controls system. These distributed simulation applications can be smaller and be distributed across multiple computing hardware units. Running faster and on demand, distributed simulation codes can reduce the overall computational load of the control system.

Simplified simulation.   The speed of simulation applications depends mainly on their algorithmic complexity and modeling resolution. Simpler models and simplified algorithms could reduce the computational load. Simplification and lower level of modeling detail could of course reduce the reliability of predictions and must be thus scrutinized on a case-by-case basis.

Simulation substitutes.   Fundamental computational functionalities of detailed simulation applications may be captured by computationally more efficient regression models or neural network copies of simulation applications. Regression models are derived based on systematic multiple runs of detailed simulation programs and the statistical processing of the results. Likewise, neural networks may be trained by data generated through multiple runs of simulation programs. The advantage of these approaches lies in the very high speed of neural network computing and regression models. Such modeling techniques obviously lack the flexibility of explicit simulation methodology, but, if properly engineered, can match the predictive power of detailed simulation algorithms. Multiple designs of hybrid control systems that utilize both simulation and machine learning have been designed and successfully tested (Chang and Mahdavi 2002).

Rules represent a further class of—rather gross—substitutes for simulation-based behavioral modeling. In certain situations, it may be simpler and more efficient to describe the behavior of a system with rules, instead of simulations. Such rules could define the relationship between the state of a device and its corresponding impact on the state of the sensor. Rules can be developed through a variety of techniques. For example, rules can rely on the knowledge and experience of the facilities manager, the measured data in the space to be controlled, or logical reasoning.

7.4.6   Case studies

7.4.6.1   Overview

To provide further insights into the problems and promises of simulation-based control strategies, we present in the following sections, two illustrative case studies involving exploratory implementations. The first case study addresses the daylight-based dimming of the electrical lighting system in a test space (Section 7.4.6.2). The second case study is concerned with the thermal control of a test space (Section 7.4.6.3).

7.4.6.2   Daylight-based dimming of the electrical light in a test space

We introduced the simulation-based control method using an illustrative case, which involved the selection of a preferable louver position toward improving the daylight availability and distribution in a test space (see Section 7.4.4). In this section, we consider the problem of daylight-based dimming of the electrical lights in the same test space (Mahdavi 2001a). The objective of this control strategy is to arrive at a configuration of daylighting and electrical lighting settings that would accommodate the desired value of one or more performance variables. The present scenario involves a five-dimensional control state space. As indicated before, the daylighting dimension is expressed in terms of the position of the external light redirection louvers. For the purpose of this case study, eight possible louver positions are considered. The electrical lighting dimensions encompass the dimming level of the four (independently controllable) luminaires in the space. It is assumed that each of the four luminaires in the test space can be at 1 of 10 possible power level states.

An attractive feature of a model-based control strategy is the diversity of the performance indicators that can be derived from simulation and thus be considered for control decision-making purposes. Furthermore, these performance indicators need not be limited to strictly visual criteria such as illuminance levels, but can also address other performance criteria such as energy use and thermal comfort. The lighting simulation application LUMINA can predict the values of the following performance indicators: average illuminance (E_m) on any actual or virtual plane in the space, uniformity of illuminance distribution on any plane in the space (U, cp. Mahdavi and Pal 1999), Glare due to daylight (DGI, cp. Hopkinson 1971), Glare due to electrical light (CGI, cp. Einhorn 1979), solar gain (Q), and electrical power consumption (C). The glare on the CRT (GCRT) is also considered and is taken as the ratio of the luminance of the screen to the background luminance. User's preference for the desired attributes of such performance variables may be communicated to the control system. Illustrative examples of preference functions for the performance variables are given in Figure 7.10.

These preference functions provide the basis for the derivation of objective functions toward the evaluation of control options. An objective function may be based on a single performance indicator, or on a weighted aggregate of two or more performance indicators. An example of such an aggregate function (UF) is given in Equation 7.6.

$$\mathrm{UF} = w_{E_m} \cdot P_{E_m} + w_U \cdot P_U + w_{DGI} \cdot P_{DGI} + w_{CGI} \cdot P_{CGI} \\ + w_{GCRT} \cdot P_{GCRT} + w_Q \cdot P_Q + w_C \cdot P_C \tag{7.6}$$

In this equation, w stands for weight, P for preference index, Em for average illuminance, U for uniformity, DGI for glare due to daylight, CGI for glare due to electrical light, GCRT for glare on CRT, Q for solar gain, and C for power consumption.

Needless to say, such weightings involve subjective and contextual considerations and may not be standardized. Rather, preference functions and the weighting mechanism could provide the user of the system with an explorative environment for

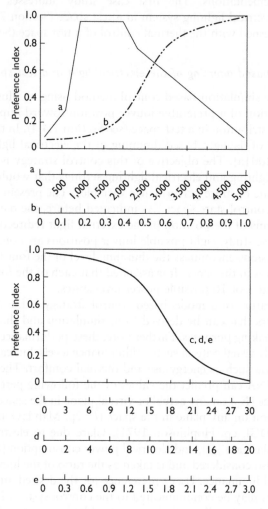

Figure 7.10 Illustrative preference functions for selected performance variables (a) Average illuminance in lx; (b) Uniformity; (c) DGI; (d) CGI; (e) GCRT.

the study of the relative implications of the impact of various performance indicators in view of preferable control strategies. To generate suitable schemes for daylight-responsive electrical lighting control, we considered two possibilities. The first possibility involves the simultaneous assessment of various combinations of the states of the daylighting and electrical lighting control devices. This strategy requires, due to the potentially unmanageable size of the resulting control state space, a reduction of the possible number of states: Let D be the number of luminaires (or luminaire groups) and P the number of dimming positions considered for each luminaire. Using Equation (7.4), the total number of resulting possible combinations (control states) can be computed. For example, for $D = 4$ and $P = 10$, a total of 1,048,576 possible electrical lighting control states results. Assuming eight daylight control states (eight louver positions), a total of 8,388,608 simulation runs would be necessary at each time-step. Detailed lighting simulation runs are computationally intensive and require considerable time. Obviously performing this number of simulations within a time-step (of, say, 15 min) is not possible. To reduce the size of the segment of the control state space to be explored, one could either couple devices (e.g. by dimming the four space luminaires in terms of two coupled pairs) or reduce the number of permissible device positions. An example for the latter would be to consider, at each time-step, only three dimming states for each luminaire, namely the status quo, the immediate higher state, and the immediate lower state. In the present case, this would mean that $D = 2$ and $P = 3$, resulting in 9 electrical lighting options. Considering 4 candidate louver positions, the total number of required simulations would be reduced to the manageable number of 36.

The concurrent simulation-based assessment of daylight and electrical light options allows for the real-time incorporation of changes in room and aperture configuration, as well as flexibility in the definition of the relevant parameter for performance variables (such as the position of observer, etc.). However, the limitation of possible dimming options at each time-step to the immediate adjacent positions may result in the inability of the search process to transcend local minima and/or maxima. This problem can be handled to a certain degree by considering additional randomly selected control state options to be simulated and evaluated in addition to the default "greedy" search option in the control state space (cp. Section 7.4.5.2).

The second approach to the generation and evaluation of alternative control options involves a sequential procedure. In this case, first, the preferable louver position is derived based on the methodology described earlier. The result is then combined with a preprocessed matrix of various luminaire power levels. This matrix (or look-up table) can be computed ahead of the real-time control operation based on the assumption that the incident electrically generated light at any point in the space may be calculated by the addition of individual contributions of each luminaire. The matrix needs only to be regenerated if there is a change either in the configuration of interior space or in the number, type, or position of the luminaires. The advantage of this approach is the possibility to reduce computational load and extend the search area in the control state space. The typical time interval between two actuation events (e.g. change of louver position and/or change of the dimming level of a luminaire) would then be generally sufficient to allow for the simulation of an increased number of louver positions. Combining the results of the selected louver settings with the matrix of electrical lighting states does not require real-time simulation and is thus efficient computationally. As a result, a larger number of dimming options may be considered and evaluated toward the selection of the preferable combined daylighting and electrical lighting settings.

The following steps illustrate this process for a generic time-step as experimentally implemented in IW:

1   Outdoor light conditions, the current louver position, luminaire power levels, and the current time were identified (Table 7.5).
2   Simulations were performed for each of the eight candidate louver positions based on the input data. Calculated performance indices for each louver position were further processed to generate the utility value (UF) based on the preference indices and corresponding weights (Table 7.6). Subsequently, the louver position that maximizes utility was selected (105°).

Table 7.5 Initial state as inputs to simulation

Year	Month	Day	Hour	I_{global} (W/m²)	$I_{diffuse}$ (W/m²)	E_{global} (lx)	θ_n (lvr) (degree)	L1 (%)	L2 (%)	L3 (%)	L4 (%)
1998	5	12	15	343	277	39,582	30	50	40	40	50

Note
L1, L2, etc. are current luminaire input power levels

Table 7.6 Performance indices and the utility values for each optional louver position

θ_{n+1} (lvr) (degree)	E_m (lx)	U_E	DGI	CGI	GCRT	Q (W)	P (W)	UF
0	291	0.849	4.31	0	0.744	4.29	0	0.623
15	249	0.856	4.14	0	0.752	3.84	0	0.593
30	251	0.855	4.28	0	0.749	3.59	0	0.594
45	263	0.870	4.39	0	0.742	3.54	0	0.606
60	280	0.859	5.56	0	0.739	3.46	0	0.617
75	310	0.430	5.81	0	0.731	3.57	0	0.665
90	331	0.840	5.98	0	0.707	3.90	0	0.665
105	337	0.841	6.00	0	0.747	4.47	0	0.670

Note
Weights: $w_{E_m} = 0.45$, $w_U = 0.2$, $w_{DGI} = 0.05$, $w_{CGI} = 0.03$, $w_{GCRT} = 0.1$, $w_Q = 0.12$, and $w_P = 0.05$.

3   Another round of simulations for the selected louver position was performed to generate intermediate data for the calculation of glare indices when the selected louver position is combined with various sets of luminaire power level configurations. Calculated glare component parameters (daylight component) include background luminance, luminance of each window patch for DGI calculation, direct and indirect illuminance on the vertical surface of the eye for CGI calculation, as well as the luminance on the computer screen for GCRT calculation.

4   For each luminaire, five steps of candidate power levels (current power level plus two steps below and two steps above) were identified. Then, from the pre-generated look-up table, all 625 (5^4) power level combinations were scanned to identify the corresponding illuminance distribution and power consumption along with the glare component parameters (electrical light component) for CGI and GCRT calculations.

5   Final values of glare indices were generated by combining the glare component parameters (both daylight component and electrical light component) calculated in step 3 and 4 for each louver–luminaire set. This is possible since the pre-calculated glare component parameters are additive in generating the final glare indices.

6   The louver position and luminaire power levels for the preferable control state were identified by selecting the one option out of all 625 sets of louver–luminaire control options that maximizes the utility value (cp. Table 7.7).

7   Analog control signals were sent to the louver controller and luminaire ballasts to update the control state.

Table 7.7 Selected control option with the corresponding performance indices and utility

θ_{n+1} (lvr) (degree)	L1 (%)	L2 (%)	L3 (%)	L4 (%)	E_m (lx)	U_E	DGI	CGI	GCRT	Q (W)	P (W)	UF
105	30	20	20	30	698	0.913	3.93	0	0.561	4.47	58	0.917

Table 7.8 Implementation parameters

Type and number of devices		States	Control parameters
Light redirection louvers	1	0°, 70°, 90°, and 105° from vertical	Illuminance, Temperature
Electric lights	8	0%, 33%, 67%, and 100%	Illuminance
Heating valve	2	0%–100% (in 5% increments)	Temperature

7.4.6.3   Thermal control of a test space

The following demonstrative implementation investigates cooperation among devices within the same building service domain, the interaction between multiple domains, and the interaction between two spaces that share the same device. The objective function of the control system is to maintain all control parameters (as monitored by sensors) within their set-point ranges while considering a number of constraints. In this case study, both simulation- and rule-based control functionalities are applied. The configuration of the spaces used in this implementation is the same as the one shown in Figure 7.6. Each space contains four electric lights and a local heating valve. An exterior light-redirection louver system is shared by both spaces. The device states have been discretized for control state space reduction, and the performance indicators impacted by each device are listed in Table 7.8.

The control system was virtually operated for four days (in the heating season during daylight hours) for which the following sensor data were available: interior illuminance and air temperature, outdoor air temperature, external (horizontal) global and diffuse irradiation, and central system hot water supply temperature.

The object model generated for this implementation is shown in Figure 7.8. For this experiment, simulation-based control methodology was implemented in nodes DC-Va1, DC-Va2, and DC-Lo1. Rule-based control methodology was used for the remaining nodes. The following example describes how rules were developed from measured data to capture the impact that the states of four electric lights had on the space interior illuminance.

The luminaire rules (implemented in the DC-EL nodes) were developed from measured data taken during night so that the influence of daylight on the results was excluded. The electric lights were individually dimmed from 100% to 0% at 10% intervals, and the desktop illuminance was measured. Figure 7.11 shows, as an example, the impact each luminaire (and its dimming) has on sensors E1. Further rules utilized at each MC node are summarized in Table 7.9.

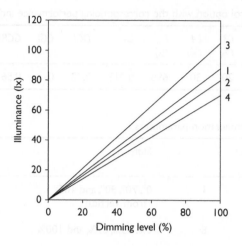

Figure 7.11 Measurements for interior illuminance rule: The impact of the dimming level of the luminaires 1, 2, 3, and 4 on the measured illuminance (for sensor E1).

Table 7.9 Rules used for implementation

Node	Rule
MC-EL_1 and MC-EL_2	Prohibit independent switching (i.e. lights dim together)
MC-EL_Lo_1 and MC-EL_Lo_2	Fully utilize daylighting before electric lighting
MC-Lo_VA_1 and MC-Lo_VA_2	Fully utilize solar heat before mechanical heating
MC-II EL_Lo_VA_1	Choose option that meets set-point need of all sensors

To implement the simulation-based control method, the Nodem energy simulation tool was used (Mahdavi and Mathew 1995; Mathew and Mahdavi 1998). Nodem predicts interior temperature by balancing heat gains and losses in the space. The local heating valve was simulated as a heat gain to the space, which was added to other internal loads in Nodem. It was necessary to determine how much heat gain to each space is possible through the water local heating system at each valve state. The local supply temperature is dependent on the central supply temperature, which changes continually due to the changing needs of the building. The heat supplied to the space is dependent on local supply temperature. Thus, the amount of heat provided by the local heating system changes with constant valve state. Estimating the losses from the mullion pipes to the space was accomplished by estimating the local water flow rate and measuring the surface temperatures at both ends of the pipe. Over the course of several days in the winter, the water mullion valve was moved to a new position every 20 min, and the resulting surface temperatures measured. The heat loss to the space was calculated for a valve position of 100% and binned according to the central system water supply temperature. The results are graphed in Figure 7.12 and provide the basis for a rule used by the DC-Va nodes to estimate the heat gain values needed for simulation.

The louvers are simulated in both LUMINA and Nodem. In LUMINA, illuminance changes due to louver position were determined by modeling the louver as an exterior surface. LUMINA calculates the inter-reflections of light between the louver surfaces as well as between the louvers and window surfaces. To calculate the amount of solar heat gain to the space at a given time, LUMINA was used as well. The resulting solar heat gain was then input into Nodem as an additional heat gain. Note that LUMINA was calibrated to provide a more accurate prediction of interior illuminance levels. This calibration was performed based on the comparison of a series of measured and simulated illuminance level in the space. Figure 7.13 illustrates the relationship between measured and simulated illuminance levels (for sensor E_1) before (B) and after (A) the calibration.

The virtual operation of the control system at each time-step begins with measured illuminance and temperature data that are mapped to the sensor representations in the object model. The device controllers read the new sensor values, determine whether they are out of range, decide on appropriate action based on their decision-making

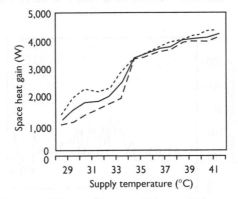

Figure 7.12 Heat output of DC-Va as a function of supply temperature.

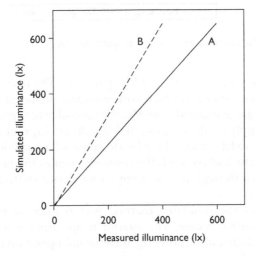

Figure 7.13 The relation between measured and simulated illuminance levels (sensor E_1) before (B) and after (A) calibration of LUMINA.

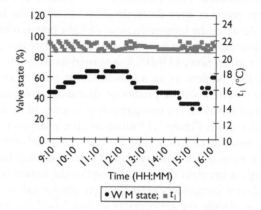

Figure 7.14 Simulated valve positions and space temperatures in the test space for Day 1.

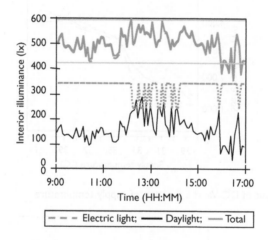

Figure 7.15 Simulated illuminance levels in the test space for Day 1.

algorithm, and submit a request to their MC parent(s). The format of the request is a table of optional device states and their corresponding predicted impact on each sensor with which they are associated. The MC is responsible for aggregating the requests of its child nodes, applying the decision-making algorithm, ranking the options, and supplying its parent node(s) with a list of state options for each device state for which it is responsible. At the highest level, the controller makes the final decision, sending the results back down through the hierarchy to the device controllers, which then set the new device states.

Figure 7.14 shows the simulated thermal performance of the test space while controlled by this control system. The interior temperature is maintained within its set-point range. The figure also shows the incremental operation of the water mullion valve in response to changes in temperature.

Figure 7.15 shows, as an example, the simulated interior illuminance in the test space for Day 1. The cooperation between electric light and daylight is apparent as the electric light component of illuminance drops (electric light state decreases) as daylight increases. The total interior illuminance is generally maintained within the set-point range (400–600 lx).

7.5 Conclusion

This chapter described the concepts of sentient buildings and self-organizing building models. It explored the possibility of integrating primary models of buildings' composition and behavior in higher-level building control systems. The realization of such a meta-mapping functionality in the building operation system architecture could extend the role and applicability of dynamic (behavioral) building representations beyond their current use in computational performance-based building design support. Computational performance simulation codes and applications could become an integral part of the methodological repertoire of advanced building operation and controls systems. Thus, a larger set of indoor environmental performance indicators could be considered toward indoor climate control. Beyond mere reactive operations based on environmental sensing, simulation-based building control allows for proactive evaluation of a richer set of control options, and allows for the reduction of the sensors needed for the real-time environmental performance assessment. A particularly attractive feature of the proposed model-based strategy lies in its potential for a transparent and high-level integration of multiple control agenda. Thus, complex control strategies may be formulated to simultaneously address habitability, sustainability, and feasibility considerations in providing appropriate levels of building performance.

References

Broz, V., Carmichael, O., Thayer, S., Osborn, J., and Hebert, M. (1999). "ARTISAN: An Integrated Scene Mapping and Object Recognition System." *American Nuclear Society* In 8th Intl. Topical Meeting on Robotics and Remote Systems, American Nuclear Society.

Chang, S. and Mahdavi, A. (2002). "A hybrid system for daylight responsive lighting control." *Journal of the Illuminating Engineering Society*, Vol. 31, No. 1, pp. 147–157.

De Ipina, D.L., Mendonca, P., and Hopper, A. (2002). "TRIP: a low-cost vision-based location system for ubiquitous computing." *Personal and Ubiquitous Computing Journal*, Springer, Vol. 6, No. 3, pp. 206–219.

Eggert, D., Fitzgibbon, A., and Fisher, R. (1998). "Simultaneous registration of multiple range views for use in reverse engineering of CAD models." *Computer Vision and Image Understanding*, Vol. 69, pp. 253–272.

Einhorn, H.D. (1979). "Discomfort glare: a formula to bridge difference." *Lighting Research & Technology*, Vol. 11, No. 2, pp. 90.

Faugeras, O., Robert, L., Laveau, S., Csurka, G., Zeller, C., Gauclin, C., and Zoghlami, I. (1998). "3-D reconstruction of urban scenes from image sequences." *Computer Vision and Image Understanding*, Vol. 69, pp. 292–309.

Finkenzeller, K. (2002). RFID-Handbuch. Hanser. ISBN 3-446-22071-2.

Hopkinson, R. G. (1971). "Glare from window." *Construction Research and Development Journal*, Vol. 2, No. 4, pp. 169–175; Vol. 3, No. 1, pp. 23–28.

Mahdavi, A. (2003). "Computational building models: theme and four variations (keynote)." In *Proceedings of the Eight International IBPSA Conference*. Eindhoven, The Netherlands. Vol. 1, ISBN 90-386-1566-3, pp. 3–17.

Mahdavi, A. (2001a). "Simulation-based control of building systems operation." *Building and Environment*, Vol. 36, Issue No. 6, ISSN 0360-1323, pp. 789–796.

Mahdavi, A. (2001b). "Aspects of self-aware buildings." *International Journal of Design Sciences and Technology*. Europia: Paris, France. Vol. 9, No. 1, ISSN 1630-7267, pp. 35–52.

Mahdavi, A. (2001c). Über "selbstbewusste" Gebäude. Österreichische Ingenieur- und Architekten-Zeitschrift (ÖIAZ), 146. Jg., Heft 5-6/2001, pp. 238–247.

Mahdavi, A. (1997a). "Toward a simulation-assisted dynamic building control strategy." In *Proceedings of the Fifth International IBPSA Conference*. Vol. I, pp. 291–294.

Mahdavi, A. (1997b). "Modeling-assisted building control." In *Proceedings of the CAAD Futures '97 Conference (the 7th International Conference on Computer Aided Architectural Design Futures)*. München, Germany. Junge, R. (ed.) pp. 219–230.

Mahdavi, A. (1999). "A comprehensive computational environment for performance based reasoning in building design and evaluation." *Automation in Construction*. Vol. 8, pp. 427–435.

Mahdavi, A. and Pal, V. (1999). "Toward an entropy-based light distribution uniformity indicator." *Journal of the Illuminating Engineering Society*, Vol. 28, No. 1, pp. 24–29.

Mahdavi, A., Suter, G., and Ries, R. (2002). "A represenation scheme for integrated building performance analysis." In *Proceedings of the 6th International Conference: Design and Decision Support Systems in Architecture*. Ellecom, The Netherlands. ISBN 90-6814-141-4, pp. 301–316.

Mahdavi, A., Brahme, R., and Gupta, S. (2001a). "Self-aware buildings: a simulation-based approach." In *Proceedings of the Seventh International Building Simulation (IBPSA) Conference*. Rio de Janeiro, Brazil. Vol. II, ISBN 85-901939-3-4, pp. 1241–1248.

Mahdavi, A., Lee, S., Brahme, R., and Mertz, K. (2001b). "Toward 'Self-aware' buildings. Advances in building informatics." In *Proceedings of the 8th Europia International Conference* (ed.: Beheshti, R.). Delft, The Netherlands. ISBN 2-909285-20-0, pp. 147–158.

Mahdavi, A., Chang, S., and Pal, V. (2000). "Exploring model-based reasoning in lighting systems control." *Journal of the Illuminating Engineering Society*. Vol. 29, No. 1, pp. 34–40.

Mahdavi, A. and Mathew, P. (1995). "Synchronous generation of homologous representation in an active, multi-aspect design environment." In *Proceedings of the Fourth International Conference of the International Building Performance Simulation Association (IBPSA)*. Madison, Wisconsin, pp. 522–528.

Mahdavi, A., Chang, S., and Pal, V. (1999a). "Simulation-based integration of contextual forces into building systems control." In *Proceedings of Building Simulation '99. Sixth International IBPSA Conference*. Kyoto, Japan. Vol. I, ISBN 4-931416-01-2, pp. 193–199.

Mahdavi, A., Ilal, M.E., Mathew, P., Ries, R., Suter, G., and Brahme, R. (1999b). "The Architecture of S2." In *Proceedings of Building Simulation '99. Sixth International IBPSA Conference*. Kyoto, Japan. Vol. III, ISBN 4-931416-03-9, pp. 1219–1226.

Mahdavi, A., Cho, D., Ries, R., Chang, S., Pal, V., Ilal, E., Lee, S., and Boonyakiat, J. (1999c). "A building performance signature for the intelligent workplace: some preliminary results." In *Proceedings of the CIB Working Commission W098 International Conference: Intelligent and Responsive Buildings*. Brugge. ISBN 90-76019-09-6, pp. 233–240.

Mathew, P. and Mahdavi, A. (1998). "High-resolution thermal modeling for computational building design assistance." Computing in Civil Engineering. In *Proceedings of the International Computing Congress, 1998 ASCE Annual Convention*. pp. 522–533.

Mertz, K. and Mahdavi, A. (2003). "A representational framework for building systems control." In *Proceedings of the Eight International IBPSA Conference*. Eindhoven, Netherlands. Vol. 2, ISBN 90-386-1566-3, pp. 871–878.

Pal, V. and Mahdavi, A. (1999). "Integrated lighting simulation within a multi-domain computational design support environment." In *Proceedings of the 1999 IESNA Annual Conference*. New Orleans. pp. 377–386.

Perez, R., Seals, R., and Michalsky, J. (1993). "All-Weather Model for Sky Luminance Distribution—Preliminary Configuration and Validation." *Solar Energy*, Vol. 50, No. 3, pp. 235–245.

Wouters, P. (1998). "Diagnostic Techniques." From Allard (ed.) *Natural Ventilation in Buildings: A Design Handbook*. James & James. ISBN 1873936729.

Chapter 8

Developments in interoperability

Godfried Augenbroe

8.1 Introduction

Building is a team effort in which many tasks have to be coordinated in a collaborative process. The aims and tools of the architect, the engineer, and many other players have to merge into a well-orchestrated design process. Because of the disparity of software tools, each specialist traditionally operates on an island of isolation until the time comes to match and patch with other members of the design team. Energy efficiency, optimal HVAC design, optimal visual, thermal, and acoustic comfort in buildings can only be accomplished by combining a variety of expert skills and tools through high bandwidth communication with designers in an inherently complex group process. What adds to the complexity is that the interacting "actors" come from separate disciplines and have different backgrounds. Adequate management of this group process must guarantee that design decisions are taken at the right moment with the participation of all involved disciplines. To accomplish this, one needs to be able to execute a wide variety of software applications rapidly and effectively. This has led to the need for "interoperability" between software applications. For the last fifteen years, a sustained research effort has been devoted to achieving this in the Architecture, Engineering, and Construction (A/E/C) industry. This chapter provides an overview of this work and discusses trends and future objectives of the area. The focus is on the role of building simulation tools and the typical interface problems that they pose. What started in the 1960s and 1970s as one-to-one "interfacing" of applications was soon realized to be non-scalable. In the 1980s therefore work started on the development of shared central building models, which would relieve the need for application-to-application interfaces, as depicted in Figure 8.1.

The development of the central model and the interfaces for each application grew into a new discipline, over time developing its own underpinning methods and tools, referred to as "product data technology" (PDT). The increased level of connectivity that could be achieved was termed interoperability, as indeed different applications would be able to "interoperate" through the shared data model, at least in principle. The physical data exchange takes place through software interfaces that perform mappings between global (neutral) and native (simulation view) representations. PDT provides the tools and methods for information modeling of products and all associated life cycle processes, with the aim to share that information within and across engineering, design, manufacturing, and maintenance disciplines. It should be realized that the building industry has characteristics that make the development of a "building

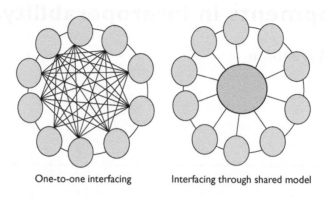

One-to-one interfacing   Interfacing through shared model

Figure 8.1 From non-scalable to scalable interoperability solutions.

product model" a huge undertaking. PDT has been efficiently deployed in highly organized engineering disciplines where it has underpinned systems for concurrent engineering, project data management, data sharing, integration, and product knowledge management. In the building industry, full-blown systems have not reached the market yet. Major obstacles are the scale and diversity of the industry and the "service nature" of the partnerships within it. The latter qualification is based on the observation that many relationships in a building project put less emphasis on predictable and mechanistic data collaboration than on the collaborative (and often unpredictable) synergy of human relationships.

The average building project requires the management of complex data exchange scenarios with a wide variety of software applications. Building assessment scenarios typically contain simulation tasks that cannot be easily automated. They require skilled modeling and engineering judgment by their performers. In such cases only a certain level of interoperability can be exploited, usually stopping short of automation. This is the rule rather than the exception during the design phases where designers call on domain experts to perform design assessments. Such settings are complex task environments where the outcome is highly reliant on self-organization of the humans in the system. The latter part of this chapter addresses the issues of integration of simulation tools in design analysis settings where these issues play a dominant role. It will be argued that interoperability according to Figure 8.1 is a requirement but by no means the solution in highly interactive, partly unstructured, and unpredictable design analysis settings. In those situations, the support of the underlying human aspects of the designer to consultant interactions, as well as the special nature of their relationship should be reflected in support systems. Different levels of integration among the team members of a building design team will have to be accomplished. First of all, there is the problem of the heterogeneity of information that is exchanged between one actor and another. Harmonizing the diversity of information in one common and consistent repository of data about the designed artifact is the first (traditional) level of ambition. The next level of integration is accomplished in the total management, coordination, and supervision over all communication that occurs within a project team.

This chapter introduces the technologies to achieve interoperability on different levels. It then reviews a number of existing approaches to develop integrated systems followed by the in-depth discussion of a new initiative in design analysis integration that combines interoperability and groupware technologies. The last section draws conclusions about the state of the art and possible next steps.

8.2 Technologies for interoperability

The previous section introduced PDT as the key-enabler of interoperability. It provides the methods and tools to develop seamless connections of software applications. The connections at the application side is typically implemented as front and backend interfaces that read/write and interpret/translate data from other (upstream) applications and produce data in a format that can be interpreted by other (downstream) applications. Achieving interoperability for building simulation applications will for instance require that design information from CAD systems can be read and automatically translated to the internal native simulation model representation, whereas the simulation outputs are in some form aggregated and automatically translated back to a neutral format which can be read by the CAD system or by other software tools such as code checking procedures, HVAC layout applications, lighting fixture design tools, or simple client report generators.

Achieving interoperability relies on the ability to identify, gather, structure, generalize, and formalize information that is exchanged between the variety of building design and engineering applications. Product models attempt to capture this information in static and generic representations. It is important to make the distinction between this product data-centric description and the information that describes the process context in which product data is exchanged. Process models capture the logic of the data generation and exchange processes that lead to the various states of the design. Hitherto, the focus of PDT is mainly on the first category of information. Process information becomes critical when one needs to manage the deployment of interoperable tools in a given scenario of use. Anticipating and coordinating the tasks in real-life scenarios require information about decision-making, design evolution and change management processes, assignment of roles and responsibilities of design actors, their possible, even their design rationale, etc. It enables the "orchestration" of the deployment of applications and other project tasks. In that case the suite of interoperable tools is embedded in a process managed interoperable system, that helps system users to execute the control over the exchange events. Section 8.2.1 deals with the data centric part of interoperability. Sections 8.2.2 and 8.2.3 then discuss the role of the process context and the technologies that are available to build integrated systems. It concludes with a brief overview of a prototype system built according to the ideas introduced in this section.

8.2.1 Data-centric interoperable systems

The area of interoperability in A/E/C has received considerable attention over the last fifteen years. An overview of projects during this period can be found in Eastman (1999). In different sectors of the A/E/C industry research and standardization initiatives were started pursuing the development of a common shared building representation. These initiatives began in the early 1990s with European Community funded research programs such as COMBINE (Augenbroe 1995) and local industry funded efforts such as RATAS (Bjork 1992), both targeting design analysis applications. An industry sector specific effort that started around the same time was CIMSTEEL, which targeted analysis, design, and manufacturing applications in the steel industry (Crowley and Watson 1997). These projects have had a major influence on the early thinking about building models and created the momentum toward efforts that followed, the most significant of which was started in 1995 by the International Alliance for Interoperability (IAI). The IAI has worldwide chapters with industrial and academic members that jointly contribute to the development of a comprehensive building model, strangely called Industrial Foundation Classes (IFC), although its aim is limited to the building industry (IAI 2002). The IFC model is an ongoing development, and although still far from complete is without doubt the most important industrial-strength landmark in AEC product modeling efforts to date. The development of a building model of the intended size of the IFC is a huge undertaking and in fact inherently unbounded unless the intended scope and usage requirements of the model are specified explicitly. Some of the issues that relate to the construction and implementation of a building product model are discussed later.

Figure 8.2 shows the example of four applications sharing information through a common representation, which will be referred to as the Building Model. The goal of the Building Model is to conceptually describe (all or a subset of) building components and abstract concepts and their relationships. Components can be defined through their compositions, functions, properties and other attributes. The choices that are faced in the definition of scope and nature of the semantic descriptions raise questions that may lead to different answers in each case. Different building models may therefore differ significantly in their structure and the abstractions that they support. Another important distinction is the way in which the modeler views the world around him, that is, as things that have an intrinsic meaning or as things that are

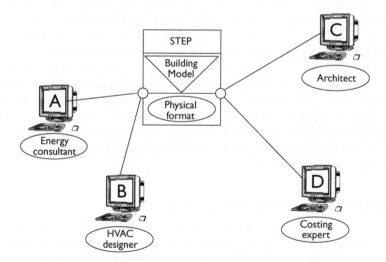

Figure 8.2 Data exchange through a central Building Model.

described solely through their functions, without attempting an intrinsic definition of things. The first approach attempts to define what things ARE whereas the second approach things are defined by what FUNCTION they perform. Depending on the viewpoint of the modeler, different models with different semantic content will result, but either model may be equally well suited to act as the facilitator of interoperability. Although the theoretical discussion is interesting (Ekholm and Fridqvist 2000), the construction of large building representations is such a steep endeavor that pragmatic choices are unavoidable. The leading IFC development is a good example of this. It should also be realized that there is no clear-cut evaluation criterion for building models, except the sobering application of the tautological rule: "if it does what it was designed for, it works."

A Building Model is typically developed as an object-oriented data model describing a building as a set of conceptual entities with attributes and relationships. Real buildings are stored as populations of this data model. The model should be complete enough so that applications can derive all their data needs from populations of the data model. Checks on data completeness by the client applications are an important part of the development process. The use of thermal building simulation program, for example, requires the derivation of those data elements that allow the construction of native geometrical representations and association of space and geometry objects to those physical properties that are necessary to perform a particular simulation. Assuming that each application's input and output data can be mapped onto this comprehensive Building Model, "full" interoperability can be accomplished. The next section will show that this statement has to be qualified somewhat, but for now it is a sufficient requirement for interoperability as implied in Figure 8.1. The figure indicates that the translation of the output of application C (e.g. a CAD system) to the neutral form as defined by the Building Model, will allow application A (e.g. a whole building energy simulation application) to extract the relevant information and map it to the native model of application A. There are different technologies that can accomplish this. A growing set of tools that deal exactly with this issue have de facto defined PDT. Especially the work by the ISO-STEP standardization community (ISO TC184/SC4 2003) has, apart from its main goal to develop a set of domain standards, added significantly to PDT. It has produced a set of separate STEP-PART standards for domain models, languages and exchange formats. The latter parts have had a major influence on the emergence of commercial toolkits for the development of interfaces. In spite of several tries (Eastman 1999), the STEP community has not been able to produce a standard Building Model. That is the reason why the industry-led initiative was launched by the IAI in 1995 with the quest to develop the IFC along a fast-track approach, avoiding the tedious and time-consuming international standardization track. After almost eight years of development, it is acknowledged that there is no such thing as fast track in building standardization (Karlen 1995a,b).

PDT-based solutions achieve interoperability by using a common formal language to express the semantics of the Building Model and an agreed exchange format for the syntax of the data transfer. The Building Model is expressed as a data schema in EXPRESS, which is the PDT modeling language of choice specified in ISO-STEP, part 11. Interfaces need to be developed for each application to map the local import/export data to the neutral Building Model. Such interfaces are described as schema-to-schema mappings and appropriate tools are available from various PDT vendors. The description of an actual building is a set of data items organized according to the Building Model schema. The actual data exchange is realized by shipping the data items in a format defined in the STEP standard, specified in ISO-STEP, part 21. This is often referred to as the STEP physical file format. Anybody familiar with XML (XML 2003) will recognize the similarity. The building representation could also be developed as an XML schema whereas actual data then would be shipped as an XML document. The application interfaces could be defined as XML schema mappings. The advancements in Enterprise Application Integration (EAI) have been largely driven by the increasing number of XML tools for this type of data transactions, and all breakthroughs in B2B applications are largely based on this technology. Although the data complexity encountered in EAI is an order of magnitude less than that is the case with multiactor integration around product models, the adoption of XML tools in PDT is growing. For the time being though, the EXPRESS language remains superior to the XML conceptual modeling tools, but the field is catching up rapidly (Lubell 2002). Current developments typically try to take the best of both worlds by doing the conceptual modeling work in EXPRESS. The resulting schema is translated into an XML schema, which then enables data instances to be sent as XML documents instead of STEP physical files. In the short run, it seems unlikely that XML technology will displace current STEP-based PDT as the latter has an advantage in operating on the scale and complexity encountered in product modeling.

The IAI has concentrated on developing its IFC as a robust, expendable, and implementable structure. Its stability and completeness has reached the stage where it can be tested in real-life settings (Bazjanac 2003), albeit in "preconditioned" scenarios. The current IFC version is a large model that has reached maturity and relative "completeness" in some domains, but it remains underdeveloped in others. A growing number of studies are testing the IFC, ascertaining its ability to provide the data that is needed by building simulation applications, for example, the data structures for an energy load analysis for building envelopes. An example of such a study is reported in (van Treeck et al. 2003).

Table 8.1 shows a typical outcome of a property matching study, in this case done as part of a graduate course assignment at Georgia Tech (Thitisawat 2003). In this case it concerns a comparison between properties needed for energy analysis applications. It was found that coverage of the IFC is very complete for most standard simulations.

In this case the material properties are taken from three IFC Material Property resources: IfcGeneralMaterialProperties and IfcHygroscopicMaterialProperties. In addition, the IFC offers a container class for user-defined properties IfcExtended MaterialProperties. This provides a mechanism to assign properties that have not (yet) been defined in the IFC specification.

The IFC model is available from the IAI and an increasing number of tool vendors have started to offer IFC interfaces (IAI 2002). Eventually, the IFC may attempt to become a STEP standard, "freezing" a version of the IFC in the form of a building industry standard.

8.2.2 The management of the data exchange

Figure 8.3 shows the realization of data exchange through transport of files between applications. Each application is expected to accept all the instances of the model and

Table 8.1 A set of IFC-offered material properties and user-added extended properties for energy
calculation purposes (Ga Tech, student project, 2003)

Property Name

IfcGeneralMaterial Properties

MolecularWeight	Molecular weight of material (typically gas), measured in g/mole.
Porosity	The void fraction of the total volume occupied by material (Vbr − Vnet)/Vbr (m^3/m^3).
MassDensity	Material mass density, usually measured in (kg/m^3).

IfcHygroscopicMaterialProperties

UpperVaporResistanceFactor	The vapor permeability relationship of air/material (typically value > 1), measured in high relative humidity (typically in 95/50% RH).
LowerVaporResistanceFactor	The vapor permeability relationship of air/material (typically value > 1), measured in low relative humidity (typically in 0/50% RH).
IsothermalMoistureCapacity	Based on water vapor density, usually measured in (m^3/kg).
VaporPermeability	Usually measured in (kg/s m Pa).
MoistureDiffusivity	Usually measured in (m^3/s).

Extended material properties: IfcExtendedMaterialProperties

	Datatype	Unit	Description
ViscosityTemperatureDerivative	REAL	kg/m-s-K	Viscosity temperature derivative.
MoistureCapacityThermalGradient	REAL	kg/kg-K	Thermal gradient coefficient for moisture capacity. Based on water vapor density.
ThermalConductivityTemperature Derivative	REAL	W/m-K2	Thermal conductivity temperature derivative.
SpecificHeatTemperature Derivative	REAL	J/kg-K2	Specific heat temperature derivative.
VisibleRefractionIndex	REAL	—	Index of refraction (visible) defines the "bending" of the solar ray in the visible spectrum when it passes from one medium into another.
SolarRefractionIndex	REAL	—	Index of refraction (solar) defines the "bending" of the solar ray when it passes from one medium into another.
GasPressure	REAL	Pa	Fill pressure (e.g. for between-pane gas fills): the pressure exerted by a mass of gas confined in a constant volume.

interpret the file according to the pre-made EXPRESS (or XML) schema, then select
those data items that are relevant for the application run that is going to be performed, mapping selected data items to the native format. Upon completion of the
operations performed by the application, the export interface maps the resulting
native data items back to the neutral model while making sure that the connections

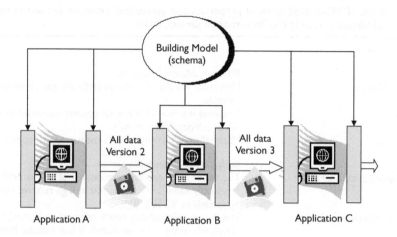

Figure 8.3 File based exchange.

with other data items in the whole model is consistently reestablished. This approach delegates the responsibility for model update and consistency management to each application, which is an undesirable side effect of this approach.

As Figure 8.3 implies, there is an import and export interface for all applications; in most cases this will be one and the same software program that is kept active while the simulation application is running. The import interface performs the selection and mapping to the native format of the application, whereas the export interface does the reverse, and "reconstructs" the links between the processed data items and the "untouched" data items in the complete model. To accomplish the latter, every software application has to understand the full structure of the Building Model. This puts a heavy burden on the interface developers, and debugging is extremely difficult because a complete set of test cases is impossible to define upfront. Another barrier to make the scenario of Figure 8.3 work is the fact that the data exchange scenario has to be choreographed in a way that the rules of data availability (at input) and data reconstruction (at output) are determinable in advance. In most cases, this will not be possible. For instance, it may not be possible to guarantee that the required data items are indeed available when the application is called. Calling the interface when not all data items are available will result in the interface module to end in an error message ("missing data"). It is usually hard to recover from this error message as it is unclear who was responsible to populate the missing items. Avoiding this situation to occur requires a certain level of preconditioning of the use-scenarios of the applications. It also has to make assumptions about the data-generation process as a whole. In-house interoperability typically allows extensive preconditioning and choreographing. Most (if not the only) successful applications of the interoperability according to Figure 8.3 have been implemented as local in-house solutions. For less predictable scenarios it is obvious that more flexible approaches are necessary.

Even with the advent of PDT toolkits and the growing set of XML-based tools, the development of an interface remains a daunting task, especially if it requires the understanding and processing of an extensive, semantically rich model. One way to

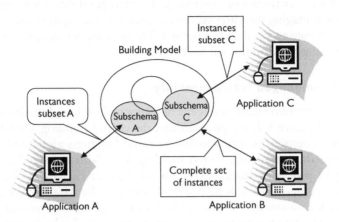

Figure 8.4 Definition of Building Model subschemas to manage the complexity of application interfaces.

relieve this necessity and make the implementation of the interfaces easier is the explicit definition of application specific subschemas. Figure 8.4 shows how this works in principle.

The definition of these subsets is not an easy task, as there exist multiple dependencies and relationships between the entities in the Building Model. For this reason the definition of subsets is all but straightforward as submodels (in fact subschemas) cannot be easily isolated from the rest of the model. Any subschema will thus have links to entities that are not part of the same subschema. These links will have to be maintained when the instances within a subschema are extracted (forming a subset of instances) and later recommitted to the populated Building Model. Conquering the overwhelming complexity of a large model through subschema definitions has only been tried occasionally, with moderate levels of success, for example, in the aforementioned projects, COMBINE and CIMSTEEL. The COMBINE subsets are driven primarily by the data needs of the applications and a set of rules has been defined to identify the "nearest" subset for an application (Lockley and Augenbroe 2000). CIMSTEEL applies the subschema approach in the same way but as a declaration (in the form of conformance classes) of meta-classes of entities that an interface is able to process. This is meant to let two applications decide, based on what conformance classes they have in common, what instance information can be exchanged. It is hard to imagine how this approach can be implemented in a file-based exchange scenario, that is, without a central data management component. In fact, it can be argued that a subschema approach requires such a central data management component to be effective and scalable. The central component will take care of the extraction and "reconstruction" task. The next section introduces a persistent database component as an essential ingredient for efficient and scalable interoperability. Efforts that have tried to implement interoperability without a central data management component are bound to fail sooner or later because of the heavy burden on the interface development and the limited scalability and testing, as well as scenario inflexibility of the solutions that are developed according to Figure 8.3. The connectivity depicted in

Figures 8.2–8.4 are examples of "unmanaged" interoperability, meaning that there is no support for managing the purpose, data availability, meaningfulness, and timeliness of the data exchange events during a project. The next section describes the issues that relate to delivering these additional functions.

8.2.3 Process-managed interoperability

Figure 8.5 shows the two major management components referred to earlier as additions to the system of Figure 8.2, wrapped around the central Building Model component. The two added components provide two essential functions already alluded to, that is, to (1) utilize the persistently stored Building Model for managed data exchange, and (2) support process-managed execution of the data exchange. The functions of the two added components will be briefly described. The database component adds persistence and data storage. It also adds data extraction and reconstruction management related to data transactions according to predefined subschemas, as explained in the previous section. In addition, all transactions are controlled and monitored by a supervision module, which will be shown to be vital to attain the desired level of system integration.

There have been few attempts to implement the supervision module of Figure 8.5. A very interesting attempt was made in the COMBINE project where it was called "Exchange Executive" (Augenbroe 1995; Amor *et al.* 1995). The module can be regarded as the component that enforces the rules that govern the pre and post conditions for the execution of a data exchange event.

The database stores the populated Building Models and supports the exchange of data to other software design tools that are integrated into the system, either by using STEP physical files or by using on-line interfaces maintained with some of the applications.

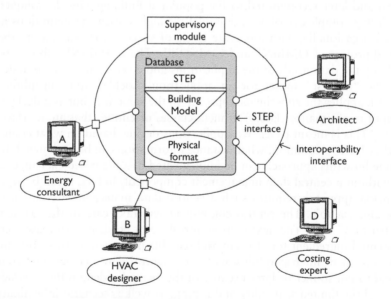

Figure 8.5 Process-driven interoperability.

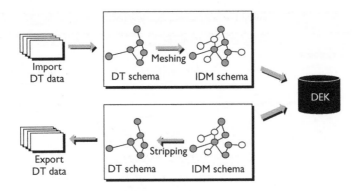

Figure 8.6 Data exchange with off-line applications.

The export and import management is handled according to predefined subschemas for each application. This requires special functionality that needs to be added to the data transaction functionality offered by commercial database systems.

Figure 8.6 shows how this was handled in the COMBINE project through a novel data exchange kernel (DEK) which performs a bidirectional data mapping between a global Building Model and the subschema of the application. The process of importing data from the application exported STEP file into the DEK is referred to as "meshing", because it merges a partial model into a richer model. Analogous to this, the process of exporting data from the DEK to an application subschema is referred to as "stripping", since it maps from a richer view of a building to a more limited view and some entity types and relationships need to be stripped off.

The primary function of the added process management component is to manage the data transaction events. Two forms of control must be regarded:

- rules and conditions determining when a specific actor can perform a particular (simulation or design) operation (this is a type of temporal control);
- rules and conditions to ensure that the Building Model instance remains in a consistent state while the design progresses (this is a type of data integrity control).

The approach that was taken in the COMBINE project implemented the two forms of control as separate components. Data integrity control was implemented as "constraint sets" in the database. The temporal control was based on the control logic embedded in a so-called "Project Window" modeling formalism (based partly on Petri-Nets). The models contain information about all communicating "actors" in the system (software applications and other tools) and their input and output schemas. In addition, they contain a formal description of the order in which these entities are allowed to execute certain operations during design evolution, that is, obeying logic dependency rules. The resulting event model formally defines the exchange event control. The "Exchange Executive" component then uses the event model to control the transitions, for example, check constraints and inform a simulation application about the data availability and the ability to perform a simulation as the next step in the process.

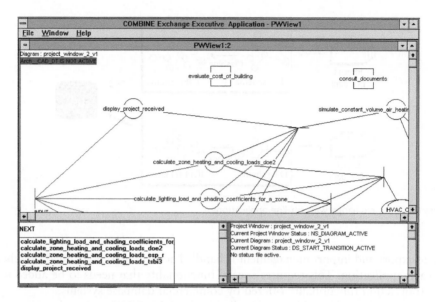

Figure 8.7 Sample ExEx startup screen (from COMBINE report).

Figure 8.7 shows a screenshot of the implemented Exchange Executive (ExEx) (Amor and Augenbroe *et al*. 1995). The event model is read by the ExEx, which is then automatically configured to assist the project manager at run time to decide which application can be executed next, and which applications are affected by changes to the Building Model and therefore need to be rerun. The state of the project moves as tokens through the network of "places" and "transitions" on the screen. It allows a team manager to decide whether a simulation software can be deployed next, based on the automatic checking of constraints.

It has been stipulated in this section that interoperability requires more than a shared Building Model and a set of applications with appropriate interfaces. Two major data and process management components have been shown to be necessary to implement interoperable solutions in practice.

The scoping of realizable and manageable systems remains an open issue. There are conflicting opinions on the size of integrated systems that can be built with current technology. Many attempts have suffered from a lack of well-defined scope. Figure 8.8 introduces a scoping approach, which has proven to be a helpful instrument in determining the size of integration efforts in general. The figure explains the difference between unscoped approaches that emphasize the generic Building Model, versus well-scoped definition of smaller Building Models in a number of separated and fairly small process windows. Each process window or "Project Window" (PW) is defined by the actors and life cycle stage that it covers. Useful PW can be identified by detecting clusters of tightly knit software applications in a particular phase of a project, involving a relatively small set of actors. Once a set of PWs have been identified, integrated systems can be targeted at them. The exchange of data across the borders of a PW is left to user-driven (traditional) procedures, involving data filtering, translation, and population of the PW-specific Building Model. These interfaces typically do not fall in the category of

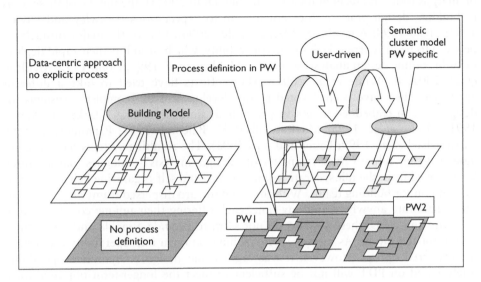

Figure 8.8 The Project Window concept.

interoperability but pose an agenda of their own. A new kind of user-driven "interface" may take the shape of a dedicated workbench that provides easy access to heterogeneous sources of information produced in other PWs.

Current approaches are aimed to deploy the IFC in a real-life process context. A recent special issue of a journal (ITCON 2003) presents working prototypes and contains papers that introduce frameworks for project and process extensions, dealing with work processes, interdisciplinary document transactions, contracting, and project management methods.

The Building Lifecycle Interoperability Software (BLIS), started in Finland is focusing on the definition of so-called "BLIS views" (BLIS 2002) uses principles that are similar to PW-based clustering. The BLIS views are process views that span a set of interoperability scenarios based on end-user "use cases". The BLIS work has delivered a set of benchmarks for the introduction of IFC in practice. Inspired by the BLIS work, the IAI development has started the use of IFC model servers to provide partial model exchange but a clear framework and architecture for process-driven communication has yet to result from these efforts.

Probably, the most important PW for interoperable simulation software applications is the design analysis PW, spanning a set of designers and domain consultants during the middle and later stages of design evolution, when they communicate design variants and design performance assessments based on simulation. Section 8.3 discusses a trial of a novel system approach particularly suited for design analysis interoperability.

8.3   Design analysis integration

This section focuses on the integration of building performance simulation tools in the building design process, achieving "true" interoperability. While a large number of analysis tools and techniques are available (DOE 2003a), the uptake of these tools in building design practice does not live up to expectations (Crawley and Lawrie 1997; Hand 1998; Augenbroe 2002; de Wilde 2004). One of the major obstacles is believed to be the current lack of interoperability, which prohibits their rapid deployment in routine design analysis scenarios (McElroy *et al.* 1997; Donn 1999; McElroy *et al.* 2001; de Wilde *et al.* 2001). Many efforts have tried to take a different approach. They have been devoted to the development of "complete" systems for designers with embedded access to (simplified) simulations (e.g. Clarke and Maver 1991; Clarke *et al.* 1995; MacRandall 1995; Balcomb 1997; Papamichael 1999; Papamichael *et al.* 1999; Baker and Yao 2002). The resulting systems have in common that they rely on fixed templates of "design analysis" dialogues that are often far removed from the idiosyncratic, spontaneous, and self-organizing behavior that is so common for building teams. With the trend toward dispersed teams of experts that can collaborate anywhere and any time, each offering their unique combination of expertise and tools, these systems do not seem to provide a viable and competitive option for the future. As a result, systems built on open interoperability are seen as the preferred option. However, it was argued that design analysis interoperability solely based on PDT will not be sufficient to meet the longer-term objective for a number of reasons, some of which have been introduced in Section 8.2. An extended discussion can be found in Augenbroe and Eastman (1998), which will be briefly summarized here:

- Current product models and standards are focused on data exchange; they do not take process context into account and therefore are unable to deal properly with data exchange control issues that are related to process logic.
- Current developments in building product models focus on single uniform ("neutral") building models. Yet neutral models have some distinct disadvantages. First, interfaces between neutral models (containing all available data about a building) and specific tools (dealing with one performance aspect only) have to filter out only the relevant information, making these interfaces overly complex ("over-engineered"). Second, the mapping of data in typical design domains to technical and performance evaluation domains (e.g. to lighting or acoustics) might not be possible. In fact, current interoperability research has failed to address the fundamental issue of the computability of this type of cross-domain mappings. Third, the use of neutral models might have implications for the order of execution of the steps in a building design process, imposing a rigid order for the use of tools and models.
- Current product models and standards assume that all information about a building design is well structured and stored in "structured idealizations" of reality. Yet, as a fact of life, a vast proportion of information will remain to live only in unstructured media such as text documents, informal memos, personal notes etc.
- Current product models assume that data mapping can be automated; this ignores that there will always be some need for additional expert-driven idealizations, based on schematization skills and engineering judgment.

These observations require the rethinking of a system approach that would be adequate for special characteristics of design analysis integration and the role of simulation tools in such a system approach. The following sections will introduce a framework for a new type of interoperability platform for building performance analysis tools in the design process.

8.3.1   A closer look at design analysis integration

Design analysis integration focuses on the effective use of existing and emerging building performance analysis tools in design analysis scenarios, with the participation of a team of designers and consultants. Some of the longer-term objectives are better functional embedding of simulation tools in the design process, increased quality control for building analysis efforts, and exploitation of the opportunities provided by the Internet. The latter refers to the possibilities for collaboration in loosely coupled teams where the execution of specific building performance analysis tasks is delegated to (remote) domain experts. It is obvious that in such teams process coordination is the critical factor with interoperability as a "support act" rather than the main objective.

Design analysis is performed through the complex interplay between design activities and analysis efforts by experts with an arsenal of simulation tools, testing procedures, expert skills, judgment, and experience. Different paradigms of expert intervention in the design processes are described in Chen (2003). The scope of our treatment of design analysis integration is limited to the assumption that the design team generates specific design analysis requests, leading to an invocation of the input of (a team of) analysis experts (Figure 8.9).

This suggests that analysis requests may be generated by a specific design activity and linked to a specific design actor responsibility. In the more generic case, the requested analysis may have ties to more than one concurrent design activity, in which case the design analysis becomes an integral part of the overall design process. In that case the analysis process cannot easily be disentangled from the complexity of other design interactions.

Figure 8.10 shows a typical situation where analysis activities become themselves an integral part of design evolution obeying design process logic while adding its own analysis logic.

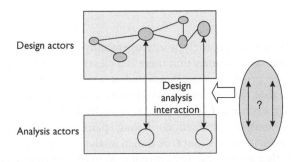

Figure 8.9 Design analysis interaction defined at specific interaction moments generated by the design team.

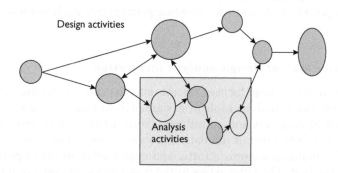

Figure 8.10 Analysis tasks with multiple interaction links with design activities.

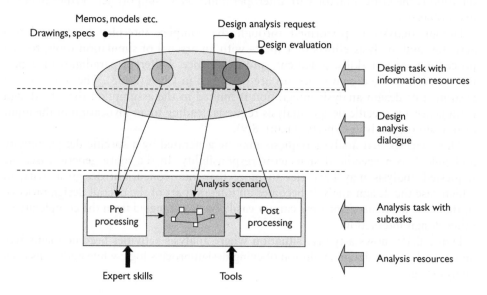

Figure 8.11 Different types of interaction and information exchange that result from a design analysis request.

A new framework for design analysis interoperability takes the situation of Figure 8.9 as starting point, that is, it assumes that there is a clearly defined design analysis request originating from one problem owner, initiated at a well-defined interaction moment. It should be noted that this situation corresponds to mainstream practice of building consultants in design evolution. The extension to multiple design analysis activities with concurrent design analysis interactions taking place will be briefly discussed in Section 8.3.5. Figure 8.11 shows that true design analysis integration faces a set of more complex dialogues that go beyond any current notion of tool interoperability.

Multiple information resources describe the state of the design at the time of the analysis request. This information is contained in different structured and unstructured documents, for example, in unstructured documents such as drawings, specifications, etc., in semi-structured documents such as CAD files, and also partly in highly structured documents such as populated IFC models. Within a design decision, an expert may be consulted upon which a design analysis request is generated in some formal manner (accompanied by a contractual agreement). Upon fulfillment of the request and completion of the analysis, an analysis report is submitted to the design team. This report is then used to perform a multi-aspect evaluation by the design team in order to fully inform a pending design decision. This evaluation may lead to the generation of new design variants for which a follow-up analysis request is issued. In many instances, a comparison of design variants may already be part of the original design request and thus part of the submitted analysis report.

As Figure 8.10 suggests, there are multiple interactions and information flows between design and analysis tasks, each of them constituting a specific element of the dialogue that needs to take place between the design team and the analysis expert. In earlier works, the main emphasis has been on support for the data connections between the design representations and the input or native models of the simulation tools. Little work has been done on the backend of the analysis. Backend integration requires the formal capture of the analysis results before they are handed back to the design team. The most relevant work in both areas is linked to the use of the IFC as structured design representation and recent work on representation of analysis results embedded in the IFC (Hitchcock *et al*. 1999).

A design analysis framework should cover all elements of the dialogue and their implementation in a configurable communication layer that drives the interoperable toolset. The following sections explain the basic constructs of the dialogue and its potential implementation such as a dialogue system.

8.3.2   A workbench for design analysis dialogues

The framework should encapsulate data interoperability in order to capitalize on efforts that have been invested in the development of building product models over the last decennium. It should not however be based on any limiting assumptions about the design process or the logic of the design analysis interaction flow. The framework should offer support for the interaction between the building design process and a wide array of building performance analysis tools. This can be realized through a "workbench" with four layers. The workbench positions building design information on the top layer and simulation applications (and more generically "analysis tools") on the bottom layers. In order to move from design information to analysis tool (pre processing) or from analysis tool to design relevant information (post processing) one has to pass through two intermediate layers. Those intermediate layers provide context to a specific interaction moment by capturing information about the process and information about the structure of the exchanged data on two separate layers, as shown in Figure 8.12. The two intermediate layers are the key layers of the workbench. They allow the domain expert to manage the dialogue between the design team (top layer) and the analysis applications (bottom layer).

Interoperability typically makes a direct connection between the top and bottom layer through a data-mapping interface. The four-layered workbench is the "fat" version of this traditional view on interoperability. The top layer contains all building

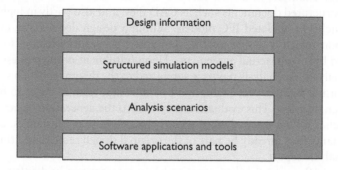

Figure 8.12 Four-layered workbench to support design analysis dialogues.

design information in partly structured and partly unstructured format. The Building Model layer contains semantic product models of varying granularity that can be used for specific analysis and engineering domains or specific performance aspects. The scenario layer captures the process logic (workflow), allowing to plan a process as well as to actually "enact" that process. These functions are offered by current mainstream workflow design and workflow enactment applications. The bottom layer contains software applications (mainly building performance simulation tools) that can be accessed from the scenario layer to perform a specific analysis. Analysis functions, rather than specific software applications are called, removing the dependency of the workbench on particular simulation software packages. This concept is fundamental to the workbench. It is based on the introduction of a set of predefined "analysis functions". Analysis functions act as the smallest functional simulation steps in the definition of analysis scenarios. Each analysis function is defined for a specific performance aspect, a specific building (sub)system and a specific measure of performance. An analysis function acts as a scoping mechanism for the information exchange between design information layer, model layer, and application layer.

The following is fundamental to the intended use of the workbench:

- The workbench is process-centric, this allows for explicit definition, management, and execution of analysis scenarios. These scenarios will typically be configured by the project manager at start-up of a new consulting job. The fact that a job can be explicitly managed and recorded offers an additional set of functions for the architectural and engineering office. Audit trails of a building analysis job can be stored and previous scenarios can be reused in new projects. This can potentially provide a learning instrument for novices in simulation. In-house office procedures and scenarios can be stored for better quality assurance.
- Expert knowledge and expertise are essential elements of performance assessment. Judgment of the applicability of performance assessment methods and evaluation of the validity of results obtained with (computational) tools are essential human skills that the workbench recognizes.

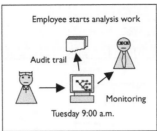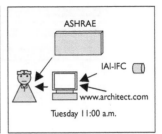

Figure 8.13 Start of the analysis process.

The following section focuses on the scenario layer and describes its central role in the workbench architecture.

8.3.3   The central role of the scenario layer

The workbench is deployed when a design interaction moment occurs and a design analysis request is triggered. The analysis expert will use the scenario layer to define the task logic of the analysis steps that respond to the request. As an example, the situation can be envisioned where an architect contacts an expert consultant in the selection of a glazing system in an office space. The expert consultant discusses the actual analysis task with the architect, plans the steps needed to carry out the analysis, and assigns one of the in-house simulation experts to carry out one or more of the analysis tasks in the scenario, as depicted in Figure 8.13.

The scenario layer typically offers access to a commercial workflow process-modeling tool to define the analysis scenarios. This type of software offers a graphical front end to a variety of "workflow enactment engines", that is, computer programs that automate the dispatching of tasks to assigned task performers, transfer of documents, coordination of dependencies between information, tasks, tools and actors in an organization etc. It allows planning a scenario that covers all steps of the analysis process, that is, from the initial design analysis request issued by the design team to the closeout of the consultancy job. It details the actual execution of the analysis, anticipates potential mid-stream modification of the analysis plan, and plans the feedback provided that is to be provided to the designer/architect. It allows graphical representation of tasks in process flow diagrams that can be constructed using drag-and-drop capabilities. It also allows easy decomposition of complex processes. A screenshot of the process-modeling window is shown in Figure 8.14, showing one of the tools available for this purpose (Cichocki *et al.* 1998).

One of the expected advantages of using a generic workflow engine is the easy integration of the workbench in environments where workflow management is already used to manage business processes. The integration of the simulation process with internal business processes such as invoicing, reporting, and resource allocation within the same (or across) collaborating firms is an exciting future prospect for the DAI workbench. It would indeed add significantly to project management and quality assurance and on the job training within the engineering enterprise.

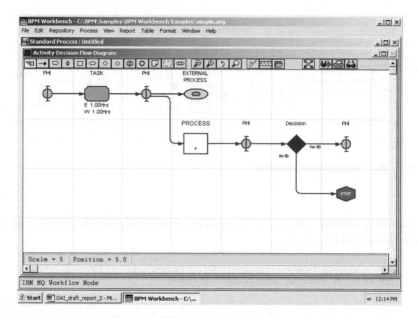

Figure 8.14 Workflow Modeling Window for analysis scenarios.

Many other useful "in-house" applications could be easily integrated in the workflow definition. For instance, keeping track of simulation results across the life cycle of a building with a tool like Metracker (Hitchcock *et al.* 1999) could be integrated into the DAI workbench.

8.3.4   The analysis function concept

Analysis functions are the key to the connection of the scenario layer with the building simulation model on one hand and the software application on the other. They allow the expert to specify exactly what needs to be analyzed and what results (captured as quantified performance indicators) need to be conveyed. To do so analysis functions need to capture the smallest analysis tasks that routinely occur in analysis scenarios. Each analysis function (AF) must identify a well-defined (virtual) experiment on the object, which is defined to reveal building behavior that is relevant to the performance aspect that is to be analyzed. The analysis function must be defined in a tool independent way, and formally specified by way of an AF-schema. The AF-schema defines the data model of the building system that the AF operates on as well as the experiment and the aggregation of behavioral output data. The analysis itself is fully embodied in the choice of an analysis function. However, not all AF calls need to be performed by a software application. For instance, some analysis function may define the daylighting performance of a window as the result of physical experiment, for example, by putting a scale model in a daylight chamber. Other analysis functions may be defined such that the measure is qualitative and subjective, based on expert engineering judgment. Allowing all these different types of analysis functions to be part of the same environment and controlled by a transparent workflow model adds to the control over the process, especially if one realizes that many different experts may be called upon in the same project, each using their own analysis expertise.

An analysis function is defined as an experiment needed to generate behavior (building states over time) that can be observed; from these observed states different measures for different aspects of the functional performance of the building (or building subsystem) can be derived. Each experiment is defined by the following elements:

- The experimental setup being observed (the "test box")
- The experimental conditions to which the setup is exposed (the "load" that is applied to the test box)
- The observation schedule that is used for observation of the generated states (the "measurement protocol" or "time series")
- The aggregation procedure; the observed states (the output of the experiments) are intrinsic to each experiment. Depending on the analysis function, there is an option to specify how the observed states are to be aggregated into a Performance Indicator.

Note that this approach is conceptual and disconnects the analysis function completely from its "incidental" software realization. For each individual analysis function, the entities and attributes that are described by the experiment are based on generalizations of performance analysis. For instance, the analysis function for the assessment of thermal comfort may be based on the decision to evaluate thermal comfort using PMV-values (Clarke 2001). Because of this, the analysis function needs to describe those entities that are needed to calculate PMV-values: there needs to be an internal air zone that has an average air temperature, and there need to be surfaces that have temperatures that can be used to calculate a mean radiant temperature. Also, occupants need to be defined that have a metabolic rate and clothing value. However, if the decision had been made to base the thermal comfort analysis function on a different measure, for instance the use of degree hours for the air temperature, then there would not have been a need to include any occupant and occupant properties, and the treatment of surfaces might have been different. Note that different analysis functions for thermal comfort, like a PMV-based and a degree hour-based function can coexist, and can be kept independent of their software realization.

Figure 8.15 shows an example of a structured description of an AF, in this case to perform the experiment for energy performance evaluation in an elementary space.

8.3.5 The DAI prototype

The framework introduced in the previous sections has been implemented in the Design Analysis Interface-Initiative (Augenbroe and de Wilde 2003; Augenbroe *et al.* 2004). Its main objective was to test the proposed framework with focus on the scenario development and the embodiment of AF-driven interfaces to existing simulation tools. This section describes the prototype development in more depth.

AF NAME:	"Analyze efficient use of energy for an office cell of type X"
VERSION:	Created December 2001; last modified may 2003

SYSTEM:

Subsystems:	Assumptions:
1. Internal air zone (3.6 x 2.7 x 5.4)	Complete mixing, no stratification, one average temperature
2. External air zone	Complete mixing, no stratification, one average temperature; all data dependent on location + climate (see TEST)
3. Internal construction elements (0.300 m of concrete)	Have thermal capacity, thermal resistance, 1-D heat flows; heat exchange with air zones through convection and radiation
4. Façade (30% glazing)	Thermal capacity, thermal resistance, 1-D heat flows; heat exchange with air zones through convection and radiation
5. Glazing system (0.006 glass, 0.012 cavity, 0.006 glass)	Thermal capacity, thermal resistance, 1-D heat flows, transmittance, reflection; heat exchange with air zones through convection and radiation
6. Furniture (2 desks, 2 chairs)	Assumed as added thermal mass to air zone only
7. HVAC-system	"Idealized" system

System boundaries:

Position:	Type of boundary:	Throughput:
a. At symmetry axis of construction elements	Adiabatic boundary condition	None
a. Enclosing external air zone	Formal boundary	Climate data, according to TEST

Internal system variables:

Head load, according to TEST; HVAC settings, according to TEST

OBSERVED STATES:

Observation period:

Duration:	Per week:	Per day:	Special notes:
365 days (one year)	Monday–Friday	8:00 a.m.–6:00 p.m.	Discard holidays etc

Observation time step:	Observed states:
Per hour	Heating load per year, cooling load per year

TEST:

HVAC and heat load settings Period:	Cooling set point:	Heating setpoint:	Int. heat load:
Monday–Friday 0:00 a.m.–8:00 a.m.	Off	10.0°C	None
Monday–Friday 8:00 a.m.–6:00 p.m.	22.5°C	19.5°C	2 persons, 2 PCs, lighting
Monday–Friday 6:00 a.m.–24:00 p.m.	Off	10°C	None
Saturday–Sunday 0:00 a.m.–24:00 p.m.	Off	10°C	None

Air exchange rate:	Climate data:
Constant, at 1.0 h^{-1}	According to TMY–2 data for Atlanta, GA, USA

AGGREGATION OF OBSERVED STATES:

a = heating load per year (observed state 1)	c = heating load per year for reference case Z
b = cooling load per year (observed state 2)	d = heating load per year for reference case Z
e = a + b	f = c + d
PI = e/f	

Figure 8.15 Structured format for AF description.

The AF Models are translated into XML schemas in order to be machine readable. Each AF in fact represents a minimal "product model" and a (small) subschema of the kind presented in Section 8.2.3. In this case the subschema does not represent the input model of a simulation tool, but the combination of input and output of a particular analysis function irrespective of the simulation tool that will perform it. Typically the AF schema is much smaller than the subschema of the complete input and/or output model. The other difference is that AF schemas are not defined as "parts" of a bigger Building Model (as in Section 8.2.3) but as bottom-up defined schemas that define a modular and tool independent analysis step, which recurs routinely in design analysis scenarios.

At run time an AF model needs to be populated through data interfaces that "pull" data from the models that reside in the building analysis model layer. The pull approach allows for a directed search of relevant information that resides on the upper layers of the workbench. Whenever building information is present in structured format, the data exchange from upper layer to the AF Model can be automated. However, the AF Model population interface may signal missing information and trigger manual input by the simulation expert. Since the AF Model only describes the elements that are critical for a small performance analysis, both the automatic part of the interface as well as the user guided constructive part represent small and manageable programming efforts for the developers. They will benefit greatly from emerging XML-based graphical (declarative) mapping languages. Each AF drives the connection and data exchange with the neighboring workbench layers. The principle of the approach is shown in Figure 8.16.

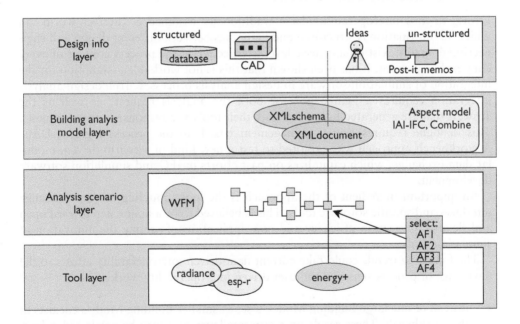

Figure 8.16 Analysis function as central driver in the data exchange topology.

Only a very rudimentary test of the workbench could be performed with three analysis functions, that is, for energy efficiency, thermal comfort, and daylight autonomy. Two simulation tools were embedded in the tool layer of the prototype to carry out these functions: EnergyPlus (DOE 2003b) for the energy and thermal comfort analysis, and IdEA-L for the quantification of the daylight autonomy (Geebelen and Neuckermans 2001).

In a full-blown implementation, the tool layer would in fact contain a great variety of simulation tools that would perform the simulation defined by an AF. Obviously only those tools can be called that have been tested to be valid for the particular AF and for which an AF mapping to input and output data has been predeveloped. The software that is actually called is controlled by the designer of the workflow, although it could be equally valid to match each AF with a software module that has been "preaccredited" for this function so that the workflow designer need not worry about the association of a needed AF with a software application. An overview of a test-run with the prototype on a very simple scenario (the selection of a window component in façade design) is described in Augenbroe and de Wilde (2003).

Future developments of the DAI-Prototype should first of all deal with the expansion of the set of analysis functions together with the population of the tool layer with additional software tools, each equipped with multiple small AF-based interfaces. The next step should then deal with the development of the constructive interfaces that populate AF models from design information. These interfaces will be small and manageable and thus be easily adaptable to the changing and growing neutral product models (such as IFC) from which they are mapped.

8.4   Conclusions and remarks

Efforts toward interoperability have traditionally assumed a "perfect world" in which all information is structured and mappings between different design and engineering domains exist on a generic level irrespective of the process context of every data exchange event. It has been shown that this is not workable assumption for the integration of simulation software in design analysis processes. True design analysis integration requires a "language" for both the analysis requests as well as the answers that are generated by experts with their tools as a response to these requests. This dialogue requires proper management based on the process logic of DAIs. A workbench approach was explored to test a new kind of scenario-driven, modular data exchange, which capitalizes on past efforts in IFC and simulation software development.

An important ingredient of the approach is the loose coupling of data exchange interfaces and specific software tools. This is believed to be a major step toward open and flexible integration of legacy and new applications, fostering the innovation of simulation tools.

The following trends could take current data-driven interoperability efforts to the next stage of process-sensitive and user-driven interoperability workbenches:

- instead of one large Building Model, a set of smaller Building Models coexists in the workbench. These reside on a separate layer, and may be partly redundant; they are domain and "process window" specific and may contain varying levels of "idealized" representations;

- the mapping between design information and the set of coexisting structured representations is supported by constructive interfaces that are integrated in the workbench. Each interface operates in pull mode, initiated by the occurrence of predefined analysis functions in scenarios. This approach avoids "over-engineered" interfaces;
- the constructive interfaces are "procedural" replacements of the integrated Building Model;
- the scenario approach integrates well with the trend to support collaborative design teams by Internet-based team.

A number of important research questions need to be addressed before large-scale interoperability workbench development can be undertaken:

1   Can a distinct set of molecular AFs be defined that covers a significant enough percentage of recurring analysis scenarios? The underlying performance theory has roots in earlier work by CSTB on Proforma (CSTB 1990) and CIB's work on test methods. Later work has tried to establish Performance Assessment Methods, such as reported in (Wijsman 1998). However, a classification of system functions and their performance measures has not been attempted at any significant scale.
2   Can the claim for maximum reuse of IFC investments be proven? The IFC is being tested successfully in preconditioned settings. A hard test for the IFC would be to develop a large enough set of analysis functions and test the coverage of the data needs of these analysis functions by the IFC. From preliminary studies, it seems that the coverage in the energy, HVAC, and related analysis fields would be complete enough to cover a significant set of analysis functions.
3   Can the workbench approach effectively capture performance at increasing levels of granularity in accordance with design evolution or will the necessary number of analysis functions explode? The answer to this question will be largely determined by the establishment of an AF classification and the way this classification can be applied to different building systems and to varying levels of granularity.
4   Will the workbench approach lead to the capturing of best practices in current building/engineering design and thus be able to act as a catalyst for reengineering? The building performance analysis profession is gaining in maturity but lacks clear standards and accepted quality assurance methods. The diffusion of best practices could prove to be an important factor for this maturation.
5   Who will own and maintain the classification of analysis functions? Does this not have the same drawback/problem as the ownership and maintenance of the neutral product model? There is as yet no way of knowing this, as the complexity of the classification is untested.

These are important questions for which no definite answers can be given as yet.

Acknowledgment

The reported work on DAI was funded by the US Department of Energy and conducted by Georgia Institute of Technology, University of Pennsylvania and Carnegie Mellon University.

References

Amor, R., Augenbroe, G., Hosking, J., Rombouts, W., and Grundy, J. (1995). "Directions in modeling environments." *Automation in Construction*, Vol. 4, pp. 173–187.

Augenbroe, G. (1995). COMBINE 2, Final Report. Commission of the European Communities, Brussels, Belgium. Available from http://dcom.arch.gatech.edu/bt/Combine/my_www/document.htm

Augenbroe, G. (2002). "Trends in building simulation." *Building and Environment*, Vol. 37, pp. 891–902.

Augenbroe, G. and de Wilde, P. (2003). "Design Analysis Interface (DAI)." Final Report. Atlanta: Georgia Institute of Technology. Available from http://dcom.arch.gatech.edu/dai/

Augenbroe, G. and Eastman, C. (1998). "Needed progress in building design product models." White paper, Georgia Institute of Technology, Atlanta, USA.

Augenbroe, G., Malkawi, A., and de Wilde, P. (2004). "A workbench for structured design analysis dialogues." *Journal of Architectural and Planning Research* (Winter).

Balcomb, J.D. (1997). "Energy-10: a design-tool computer program." Building Simulation '97, 5th International IBPSA Conference, Prague, Czech Republic, September 8–10, Vol. I, pp. 49–56.

Baker, N. and Yao, R. (2002). "LT Europe—an integrated energy design tool. Design with the Environment." In *Proceedings of the 19th International PLEA Conference*. Toulouse, France, July 22–24, pp. 119–124.

Bazjanac, V. (2003). "Improving building energy performance simulation with software inter-operability." Building Simulation '03, 8th International IBPSA Conference, Eindhoven, The Netherlands, August 11–14, pp. 87–92.

BLIS (2002). Building Lifecycle Interoperable Software Website. Available from http://www.blis-project.org/

Bjork, B.-C. (1992). "A conceptual model of spaces, space boundaries and enclosing structures." *Automation in Construction*, Vol. 1, No. 3, pp. 193–214.

Chen, N.Y. (2003). "Approaches to design collaboration research." *Automation in Construction*, Vol. 12, No. 6, pp. 715–723.

Cichocki, A., Helal, A., Rusinkiewicz, M., and Woelk, D. (1998). *Workflow and Process Automation, Concepts and Technology*. Kluwer, Amsterdam.

Clarke, J.A. (2001). *Energy Simulation in Building Design*. Butterworth Heinemann, Oxford.

Clarke, J.A. and Maver, T.W. (1991). "Advanced design tools for energy conscious building design: development and dissemination." *Building and Environment*, Vol. 26, No. 1, pp. 25–34.

Clarke, J.A., Hand, J.W., Mac Randal, D.F., and Strachan, P. (1995). "The development of an intelligent, integrated building design system within the European COMBINE project." Building Simulation '95, 4th International IBPSA Conference, Madison, WI, USA, August 14–16, pp. 444–453.

Crawley, D.B. and Lawrie, L.K. (1997). "What next for building energy simulation—a glimpse of the future." Building Simulation '97, 5th International IBPSA Conference, Prague, Czech Republic, September 8–10, Vol. II, pp. 395–402.

Crowley, A. and Watson, A. (1997). "Representing engineering information for constructional steelwork." *Microcomputers in Civil Engineering*, Vol. 12, pp. 69–81.

CSTB (1990). Informal communications from Dr. Louis Laret about standardized format to capture best practices of building analysis. CSTB, Sophia Antipolis, France.

de Wilde, P. (2004). Computational support for the selection of energy saving building components. PhD thesis. Delft University of Technology, Delft, The Netherlands.

de Wilde, P., van der Voorden, M., Brouwer, J., Augenbroe, G., and Kaan, H. (2001). "Assessment of the need for computational support in energy-efficient design projects in The

Netherlands." Building Simulation '01, 7th International IBPSA Conference, Rio de Janeiro, Brazil, August 13–15, pp. 513–520.

DOE (2003a). *US Department of Energy. Building Energy Software Tool Directory.* Washington, USA. Available from: http://www.eren.doe.gov/buildings/tools_directory/

DOE (2003b) *US Department of Energy. EnergyPlus.* Washington, USA. Available from: http://www.eren.doe.gov/buildings/energy_tools/energyplus/

Donn, M. (1999). "Quality assurance—simulation and the real world." Building Simulation '99, 6th International IBPSA Conference, Kyoto, Japan, September 13–15, pp. 1139–1146.

Eastman, C.M. (1999). *Building Product Models: Computer Environments Supporting Design and Construction.* CRC Press, Boca Raton, FL, USA.

Ekholm, A. and Fridqvist, S. (2000). "A concept of space for building classification, product modelling and design." *Automation in Construction,* Vol. 9, No. 3, pp. 315–328.

Geebelen, B. and Neuckermans, H. (2001). "Natural-lighting design in architecture: filling in the blanks." Building Simulation '01, 7th International IBPSA Conference, Rio de Janeiro, Brazil, August 13–15, pp. 1207–1213.

Hand, J.W. (1998). Removing barriers to the use of simulation in the building design professions. PhD thesis. University of Srathclyde, Department of Mechanical Engineering, Energy Systems Research Unit, Glasgow, UK.

Hitchcock, R.J., Piette, M.A., and Selkowitz, S.E. (1999). "A building life-cycle information system for tracking building performance metrics." In *Proceedings of the 8th International Conference on Durability of Building Materials and Components,* May 30 to June 3, Vancouver, BC.

IAI, International Alliance for Interoperability, (2002). Industry Foundation Classes, Release 2.2x. IAI web site, http://www.iai-international.org/iai_international/

ISO TC184/SC4 (1989–2003), ISO-STEP part standards.

ITCON (2003). Vol. 8, Special Issue IFC—Product models for the AEC Arena, http://www.itcon.org/

Karlen, I. (1995a). "Construction Integration—from the past to the present." In P. Brandon and M. Betts (eds) *Integrated Construction Information,* SPON, pp. 137–147.

Karlen, I. (1995b). "Construction Integration—from the present to the past." In P. Brandon and M. Betts (eds), *Integrated Construction Information,* SPON, pp. 148–158.

Lockley, S. and Augenbroe, G. (2000). "Data integration with partial exchange." *International Journal of Construction Information Technology,* Vol. 6, No. 2, pp. 47–58.

Lubell, J. (2002). "From model to markup: XML representation of product data." In *Proceedings of the XML 2002 Conference,* Baltimore, MD, December.

Mac Randal, D. (1995). "Integrating design tools into the design process. Building Simulation '95, 4th International IBPSA Conference, Madison, WI, USA, August 14–16, pp. 454–457.

McElroy, L.B., Hand, J.W., and Strachan, P.A. (1997). "Experience from a design advice service using simulation." Building Simulation '97, 5th International IBPSA Conference, Prague, Czech Republic, September 8–10, Vol. I, pp. 19–26.

McElroy, L.B., Clarke, J.A., Hand, J.W., and Macdonald, I. (2001). "Delivering simulation to the profession: the next stage?" Building Simulation '01, 7th International IBPSA Conference, Rio de Janeiro, Brazil, August 13–15, pp. 831–840.

Papamichael, K. (1999). "Application of information technologies in building design decisions." *Building Research & Information,* Vol. 27, No. 1, pp. 20–34.

Papamichael, K., Chauvet, H., La Porta, J., and Dandridge, R. (1999). "Product modeling for computer-aided decision-making." *Automation in Construction,* Vol. 8, January.

Thitisawat, M. (2003). IFC exam question report. Internal report, Georgia Institute of Technology.

van Treeck, C., Romberg, R., and Rank, E. (2003). "Simulation Based on the Product Model Standard IFC." Building Simulation '03, 8th International IBPSA Conference, Eindhoven, The Netherlands, pp.1293–1300.

Wijsman, A. (1998). *Performance Assessment Method (PAM) for the Dutch Building Thermal Performance Program VA114*. Delft: TNO Building and Construction Research.

XML (2003). Standard, http://www.w3.org/XML/

Chapter 9

Immersive building simulation

Ali M. Malkawi

9.1 Introduction

Advancements in visualization led to new developments in simulation. Different technologies made it possible to create environments that are virtual or augmented. Immersive simulation is the representation of the behavior or characteristics of the physical environment through the use of computer-generated environment with and within which people can interact. These environments support variety of applications including building simulation. The advantage of this simulation is that it can immerse people in an environment that would normally be unavailable due to cost, safety, or perception restrictions. It offers users immersion, navigation, and manipulation. Sensors attached to the participant (e.g. gloves, bodysuit, footwear) pass on his or her movements to the computer, which changes the graphics accordingly to give the participant the feeling of movement through the scene.

This chapter introduces a newly defined area of research we termed "immersive building simulation" and discusses the different techniques available and their applications. It illustrates how this emerging area is benefiting from the more established immersive simulation field. It begins by describing its background, which is rooted in virtual and augmented reality, and describes the application of these techniques used in different fields. It defines the essential components of immersive building simulation with a focus on data representation and interaction regarding building performance. Two example cases and recent work in this area will be discussed.

In this chapter, the term immersive building simulation is used to illustrate a specific type of simulation that uses immersive virtual or augmented reality environments. Although the term virtual reality (VR) was originally coined by Jaron Lanier, the founder of VPL Research in 1985, the concept of virtual reality can be linked to the development of calculating machines and mechanical devices (Schroeder 1993). Its modern roots can be associated with Edwin Link who developed the flight simulator in order to reduce pilot training time and cost in the early 1940s. The early 1960s show milestone developments in the field of virtual and augmented environments. In 1965, Ivan Sutherland published a paper "The Ultimate Display" (Sutherland 1965) in which he provided an argument to utilize the computer screen as a window through which one beholds a virtual world. The same year, Sutherland built the first see-through head mounted display and used it to show a wire frame cube overlaid on the real world, thereby creating the first Augmented Reality Interface. In addition, video mapping was introduced around the same time in a publication written by Myron Krueger in the early 1960s (Krueger 1985).

Table 9.1 Development of virtual environments

1940	Link Aviation developed the first flight simulator.
1957	M.L.Heilig patented a pair of head-mounted goggles fitted with two color TV units.
1965	Ivan Sutherland published "The Ultimate Display".
1971	Redifon Ltd (UK) began manufacturing flight simulators with computer graphics display.
1977	Dan Sandin and Richard Sayre invented a bend-sensing glove.
1982	Thomas Zimmerman patented a data input glove based upon optical sensors, such that internal refraction could be correlated with finger flexion and extension.
1983	Mark Callahan built a see-through HMD at MIT.
	Myron Krueger published "Artificial Reality".
1985	VPL Research, Inc. was founded.
	Mike McGreevy and Jim Humphries built a HMD from monochrome LCD pocket TV displays.
	Jaron Lanier, CEO of VPL, coined the term "virtual reality".
1989	VPL Research and Autodesk introduced commercial HMDs.
1992	CAVE built by University of Illinois, Chicago.
1994	Milgram introduced the term—"Mixed Reality".
2000 +	New human–computer interface mechanisms, sensors and displays for virtual environments.

It took an additional 20 years before VR hardware became relatively affordable. This is due to the introduction of PCs and the increased speed and quality of the graphics and rendering technology (Krueger 1991). In the 1980s, many commercial companies emerged to support the hardware and software of VR. These include Virtual Research, Ascension, Fakespace, etc. In addition, large industrial entities made substantial investment in the technology (Boeing, General Motors, Chrysler, etc.) and academic institutions become a driving force in this development. Table 9.1 provides a summary historical view.

Virtual and augmented systems today embody a growing area of research and applications related to Human–Computer Interface (HCI). It has demonstrated tremendous benefits in many areas including commerce and entertainment. Immersive virtual reality (IVR) has only recently started to mature. Its techniques are used in the industry for product development, data exploration, mission planning, and training. Augmented Reality (AR) is still in the prototype stage, however research systems for medical, engineering, and mobile applications are now being tested. Stable hardware and graphics application programming interfaces, such as OpenGL and Performer, and reasonably priced software resulted in emerging successes for VR and AR research and applications. The Building Simulation field has been slow in taking advantage of these recent developments. Challenges related to utilizing the technology in this field will be illustrated later in this chapter. In order to discuss immersive building simulation, the concepts of immersive environments, which form the base for such simulation, will be discussed.

9.1.1 *Immersive environments*

Virtual, immersive, or synthetic environments (VEs) are computer-generated three-dimensional environments that can be interactively experienced and manipulated by the user in real-time. One way to classify immersive environments is by their end use.

There are two types of immersion systems; complete and partial immersion systems. The complete immersion leads to the virtual systems and the partial lead to augmented systems. Both systems are subclassified further by the techniques that are used to facilitate the synthetic environment and by the type of visualization they support (single user or collaborative environment).

Four key elements define this environment. First, it contains a computer-generated three-dimensional scene which requires high performance graphics to provide an adequate level of realism. The second is that the environment is interactive. Real-time response from the system is required for interaction in an effective manner. Third, the environment allows the user to be completely or partially immersed in an artificial world. Finally, the environment is registered in the three dimensions to allow this immersion to appear real. It must accurately sense how the user is moving and determine what effect this will have on the scene being rendered.

As discussed earlier, immersive environments can be subdivided into two: virtual and augmented. A visible difference between virtual and augmented environments is the immersiveness of the system. Virtual Environments strive for totally immersive systems. The visual and, in some systems, aural and proprioceptive senses are under the control of the system. In contrast, an Augmented Reality system is augmenting the real world scene necessitating that the user maintains a sense of presence in that world. The virtual images are merged with the real view to create the augmented display. Augmented Reality does not simply mean the superimposition of a graphic object over a real world scene. This is technically an easy task. One difficulty in augmenting reality is the need to maintain accurate registration of the virtual objects with the real world image. This often requires detailed knowledge of the relationship between the frames of reference for the real world, the objects to be viewed and the user. Errors in this registration will prevent the user from seeing the real and virtual images as one. The correct registration must also be maintained while the user moves about within the real environment. Discrepancies or changes in the apparent registration will range from distracting, which makes working with the augmented view more difficult, to physically disturbing for the user—making the system completely unusable.

An immersive system must maintain registration so that changes in the rendered scene match with the perceptions of the user. The phenomenon of visual capture gives the vision system a stronger influence in our perception (Welch 1978). This will allow a user to accept or adjust to a visual stimulus overriding the discrepancies with input from sensory systems. In contrast, errors of misregistration in an Augmented Reality system are between two visual stimuli, which we are trying to fuse to see as one scene. We are more sensitive to these errors in these systems, which are different from the vision-kinesthetic errors that might result in a standard virtual reality system (Azuma 1993; Azuma and Bishop 1995).

In a typical augmented system, a user wears a helmet with a semi-transparent visor that projects computer-generated images and augments the visual perception of the real environment. This is in contrast to virtual reality in which the user is completely immersed in an artificial world and cut off from the real world. Using AR technology, users can thus interact with a mixed virtual and real world in a natural way. AR can then be thought of as the middle ground between the virtual environment (completely synthetic) and the telepresence (completely real) (Milgram and Kishino 1994; Milgram *et al.* 1994; Azuma 1997; Azuma *et al.* 2001).

9.2   Immersive building simulation

An extensive body of research exists in advancing the techniques used to create immersive environments, such as calibration, mixing, integration, collaboration, etc. Although most of the work related to immersive simulation is conducted in fields that are not related to buildings, some of this work has a direct parallel and can be used to advance the work in immersive building simulation, such as immersive scientific visualization (van Dam *et al.* 2000), immersive collaborative visualization (Reed *et al.* 1997; Fuhrmann and Loffelmann 1998) and immersive real-time fluid simulation (Chen *et al.* 1997; Giallorenzo and Banerjee 1999).

In the area of engineering applications, immersive simulation is more advanced. In these applications, immersive simulation can be grouped into testing, prototyping, robotics and tele-immersion. For testing, the actual environments are simulated in a virtual world to enable engineers to interactively visualize, perceive and explore the complicated structure of engineering objects. Applications range from wind tunnel testing to Computational Steering. The virtual wind tunnel project was developed by NASA for exploring numerically generated, three-dimensional, unsteady flow fields. A boom-mounted six-degree-of-freedom and head position-sensitive stereo CRT (Cathode Ray Tube) system is used for viewing. A hand-sensitive glove controller is used for injecting tracers into the virtual flow field. A multiprocessor graphics work-station is used for computational and rendering. Computational Steering is an additional application related to immersive simulation that allows interactive control of running simulation during the execution process. Users can control a set of parameters of an operation and react to the results without having to wait for the execution process to end.

Prototyping has been used in many areas to give users a better understanding of future products and to aid in their design. It can be used in the design stage by allowing the object to be shared through distributed collaboration. It is also used for end product visualization. The advantage of the technology in this area is that once the object is erected, the technology allows for rapid changes and manipulation of its attributes while the user is fully or partially immersed.

In Robotics, the control and manipulation of robotic elements can be integrated with immersive systems. The systems allow the user to be immersed in the robot's task by using VR peripherals. The manipulation of the robotics elements such as the arm is established directly using VR devices. This technology has been widely used in telesurgery, space explorations, hazardous areas, and underwater exploration.

Tele-immersion, on the other hand, enables users at geographically distributed sites to collaborate in real-time in shared simulated environments. It is the combination of networking and media technologies that enhances collaborative environments. The concept is to recognize the presence and movement of individuals and objects, track these individuals as images and then permit them to project in immersive environments. The technological challenges include incorporation of measured and on-site data into the computational model, real-time transmission of the data from the computational model to the virtual environment and management of the collaborative interaction between two or more stations.

Although Immersive Building Simulation is still in its research and development stage and its potential application is under-researched, Virtual and Augmented

Environments have been used in a variety of areas in relation to buildings. This includes the extension of visual perception by enabling the user to see through or into objects (Klinker *et al.* 1998) such as maintenance support for visualizing electrical wires in a wall or construction grids (Retik *et al.* 1998). Other applications include structural system visualization (Fiener *et al.* 1995), augmented outdoor visualization (Berger *et al.* 1999), collaborative design process (Frost and Warren 2000), and client servicing (Neil 1996). The visualization of some of these applications becomes more useful when these environments are associated with other techniques that increase their efficiency such as knowledge-based systems (Stalker and Smith 1998).

For immersive building simulation, only a few projects have been developed—some of which are related to the post-processing of Computational Fluid Dynamics (CFD) data (Shahnawaz *et al.* 1999; Malkawi and Primikiri 2002) augmented simulations (Malkawi and Choudhary 1999); building and data representation (Pilgrim *et al.* 2001); building performance visualization (Linden *et al.* 2001; Malkawi and Choudhary 1999) and immersive visualization for structural analysis (Rangaraju and Tek 2001; Impelluso 1996). Most of the available tools provide a one- or two-dimensional representation of the data derived from a building performance simulation. This has always been an important challenge as only experts can precisely understand the data and hence are always required to interpret them. Consequently, this introduces the problems of time and cost, not only in terms of hiring these experts, but also in establishing communication among the participants. This communication is not only dependent on their physical presence. It also involves issues of representation as well as of semantics.

Immersive building simulation requires specialty hardware and software (Figure 9.1). The hardware includes the display, tracking and interaction devices. For display, immersive simulation requires a Head-Mounted Display (HMD), a Binocular Omni-Orientation Monitor (BOOM), or other peripheral hardware that allow user interaction and perception to be altered using the synthetic environment. The HMD is a helmet or partial helmet that holds the visual and auditory systems. Other immersive systems use multiple projection displays to create a room that will allow many users to interact in the virtual world. Several technologies are used for the tracking—such as mechanical, electromagnetic, ultrasonic, inertial, and optical. The interaction can be

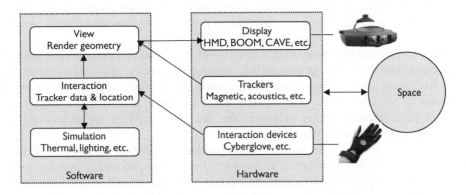

Figure 9.1 Immersive building simulation—hardware and software dependencies.

established using a data glove or 3D mouse. In addition, immersive simulation requires a three-dimensional representation of an environment and data output from a simulation.

Some of the challenges that exist in developing immersive simulation environments include the range of knowledge required in bringing together hardware, graphical and analysis software, understanding the data structure that is required for visualization and the type of interaction that is needed. Current libraries of graphical software environments provide low-level graphical operations. Some work has been done to establish mechanism for interactions to simplify the process of developing virtual environment and high-level interface between the VR graphic libraries and the analysis software (Rangaraju and Tek 2001). Most of the work is analysis specific and not widely available.

Immersive building simulation as being described here allows the user to invoke a simulation, interact with it and visualize its behavior. Its basic structure involves a simulation engine, a visualizor and hardware that allow the interaction. Although simulation engines are domain specific their output is typically data intensive and frequently requires real-time interaction. As a result, data visualization and interaction issues are the main components of this simulation. To describe immersive building simulation components, the next section will present data representation and visualization with a focus on fluids as an example. This will be followed by a discussion on the issue of interaction, which will lead to the description of the hardware and techniques used. Current research will be presented.

9.2.1   *Data visualization for immersive building simulation*

A crucial component of immersive simulation is transforming the often complex data into geometric data for the purpose of visualizing it in an intuitive and compressive way. This transformation requires the use of computer graphics to create visual images. This transformation aids in the understanding of the massive numerical representations, or data sets, which is defined as scientific visualization. The formal definition of scientific visualization (McCormick *et al.* 1987) is:

> *Scientific visualization is a method of computing. It transforms the symbolic into the geometric, enabling researchers to observe their simulations and computations. Scientific visualization offers a method for seeing the unseen. It enriches the process of scientific discovery and fosters profound and unexpected insights.*

Visual perception of the environment has always been a strong determinant in building studies. Inevitably, image processing, computer graphics, and CAD applications have had a significant impact on design methods. While computer vision assists designers in visualizing the desired built form or space, scientific visualization of abstract phenomenon further enhances the designer's capacity to understand and realize the implications of design decisions.

The visualization process in immersive building simulation consists of different stages, Figure 9.2. If visualization is considered as a post-processing operation, data generation is not part of the visualization process. On the other hand, if visualization and simulation are closely integrated, data generation is the first state of the visualization process. Regardless of the integration issue between visualization and simulation,

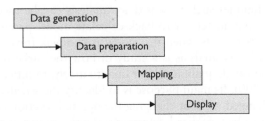

Figure 9.2 Immersive simulation—visualization process.

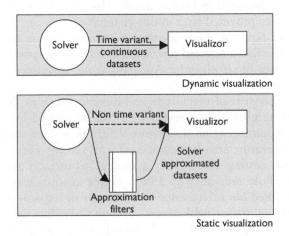

Figure 9.3 Dynamic and static data visualization.

immersive building simulation requires visualization of post-processed data. This visualization can be categorized into two types: dynamic and static, Figure 9.3.

Dynamic visualization allows continuous data connection to the visualizor. The solver speed that is responsible for delivering the raw data to the visualizor determines the success of this post-processing process. Different techniques are typically employed in order to simplify the complexity of the data and reduce computational expense, which increase the speed of the process. Such techniques include selective visualization and feature extraction. In selective visualization, a subset from a dataset that can relatively represent the actual dataset with less graphic cost is used (Walsum 1995). It uses a process of selection that is "interesting" or needed which is determined by the application area and phenomena studied and personal approach of the user. Feature extraction, on the other hand, is a process that emphasizes the selection of the data to be related to a feature (Walsum *et al.* 1996). This technique has been adapted by several leading CFD visualization programs (Kenwright 2000). Feature extraction algorithms are programmed with domain-specific knowledge and therefore do not require human intervention.

Static visualization uses data in which the time variant aspect is not directly involved. In this type, design parameters can be changed and the solver approximates the effects of the changes and creates new datasets that is directed to the visualizor. Approximation techniques such as sensitivity analysis can be used to identify crucial regions in the input parameters or to back-calculate model parameters from the outcome of physical experiments when the model is complex (reverse sensitivity) (Chen and Ho 1994). Sensitivity analysis is a study of how the variation in the output of a model can be apportioned, qualitatively or quantitatively, to different sources of variation (Saltelli *et al.* 2000). Its main purpose is to identify the sensitive parameters whose values cannot be changed without changing the optimal solution (Hillier *et al.* 1995).

The type of data to be visualized determines the techniques that should be used to provide the best interaction with it in an immersive environment. To illustrate this further, fluids, which constitute one component of building simulation, will be used as an example and discussed in further detail. This will also serve as a background for the example cases provided later in this chapter.

The typical data types of fluids are scalar quantity, vector quantity, and tensor quantity. Scalar describes a selected physical quantity that consists of magnitude also referred to as a tensor of zeroth order. Assigning scalar quantities to space can create a scalar field, Figure 9.4.

Vector quantity consists of magnitude and direction such as flow velocity, and is a tensor of order number one. A vector field can be created by assigning a vector (direction) to each point (magnitude), Figure 9.5. In flow data, scalar quantities such as pressure or temperature are often associated with velocity vector fields, and can be visualized using ray-casting with opacity mapping, iso-surfaces and gradient shading (Pagendarm 1993; Post and van Wijk 1994). Vector quantity can be associated with location that is defined for a data value in a space or in a space–time frame.

A three-dimensional tensor field consists of nine scalar functions of position (Post and van Wijk 1994). Tensor fields can be visualized at a single point as an icon or glyph (Haber and McNabb 1990; Geiben and Rumpf 1992; De Leeuw and van Wijk 1993) or along characteristic lines wherein the lines are tangent to one of the eigenvectors. The Jacobian flow field tensor visualization consists of velocity, acceleration, curvature, rotation, shear, and convergence/divergence components (de Leeuw and van Wijk 1993) (Figure 9.6).

These data types can be subdivided into two-dimensional and three-dimensional. Examples of the 2D types of the scalar quantities are iso-contours, pseudo-colors, and

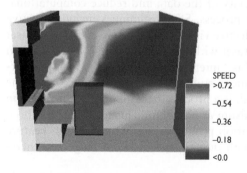

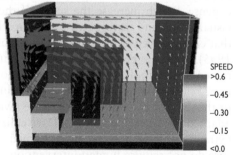

Figure 9.4 Representation of Scalar data. (See Plate VII.)

Figure 9.5 Representation of Vector data. (See Plate VIII.)

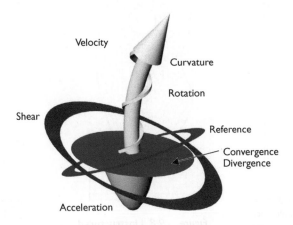

Velocity

Curvature

Rotation

Shear

Reference

Convergence
Divergence

Acceleration

Figure 9.6 Tensor—local flow field visualization. (See Plate IX.)

Table 9.2 CFD data types, visualization techniques and spatial domain

Order of data	Spatial domain	Visualization technique
Scalar	Volume	Volume ray casting
Scalar	Surface	Iso-surface
Vector	Point	Arrow plot
Vector	Surface	Stream surface
Vector	Point (space–time)	Particle animation
Tensor	Line (space–time)	Hyperstreamlines

height maps and for the vector display are arrow plots and streamlines or particle paths. In addition, these data are classified further based on the spatial domain dimensionality of the visual objects (such as points, lines, surfaces and volumes) and their association with visualization techniques (Hesselink *et al.* 1994) (Table 9.2).

In three-dimensional visualization of nonimmersive environments, the techniques known from the two-dimensional visualization of fluids are not applicable and can be misleading. However, in immersive environments, three-dimensional visualization of fluid data can be confusing. The use of iso-contours and pseudo-coloring provides a good way to visualize surfaces in the three-dimensional domain. This requires the application of some data reduction processes, such as the techniques described earlier in order to create surfaces in three-dimensional space for which two-dimensional visualization can be useful.

Besides the data type, the structure of how the data is being organized plays an important role for visualization. In fluid visualization, data is organized using computational meshes. These meshes define certain ordering of the location in space and time where the governing partial differential equations of the problem are solved numerically. All flow data is stored at these discrete locations. The data may be stored at the node of the mesh or the center of the cell of the mesh. The quality of mesh and the choice of mesh arrangement impact the accuracy of the simulation. For example,

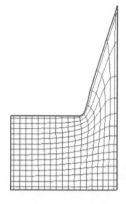

Figure 9.7 Structured grid showing regular connectivity (Courtesy: Sun 2003). (See Plate X.)

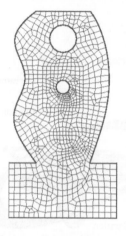

Figure 9.8 Unstructured grid showing irregular connectivity (Courtesy: Sun 2003). (See Plate XI.)

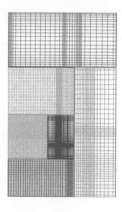

Figure 9.9 Multi-block grid showing subdomains or blocks (Courtesy: Sun 2003). (See Plate XII.)

the higher the mesh density, the higher the level of simulation accuracy, which leads to expensive computation. Five major data structures of meshes or grids are typically used: (1) structured, (2) unstructured, (3) multi-block or block-structured, (4) hybrid, and (5) Cartesian.

Structured grids consist of regular connectivity where the points of the grid can be indexed by two indices in 2D and three indices in 3D, etc. Regular connectivity is provided by identification of adjacent nodes, Figure 9.7.

Unstructured grids consist of irregular connectivity wherein each point has different neighbors and their connectivity is not trivial. In unstructured grids, the nodes and their complex connectivity matrix define the irregular forms, Figure 9.8.

Multi-block approach can break complicated geometry into subdomains or blocks; structured grids are then generated and governing equations can be solved within each block independently. These grids can handle complex geometrical shapes and simultaneously undertake a wide range of numerical analysis (Figure 9.9). Some of the advantages of multi-block grids are as follows: (a) geometry complexity can be greatly reduced by breaking the physical domain into blocks; (b) since the gridline across the blocks are discontinuous, more freedom in local grid refinement is possible; (c) standard structured flow solvers can be used within each block obviating the need for complicated data structure, book keeping and complex algorithms; and (d) provides a natural routine for parallel computing, thereby accelerating the simulation.

Hybrid grids are a combination of structured and unstructured grids and are used for better results and greater accuracy. High grid density can be designed at locations where there are sharp features of the boundary (Figure 9.10).

Cartesian grids are mostly hexahedral (right parallelopipeds) and the description of the surface is no longer needed to resolve both the flow and the local geometry

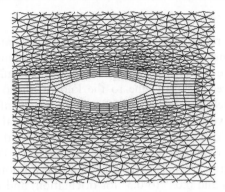

Figure 9.10 Hybrid grid—a combination of structured and unstructured grids (Courtesy: Oliver 2003).

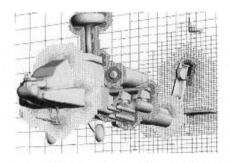

Figure 9.11 Cartesian grid (Courtesy: Aftosmis *et al.* 1998). (See Plate XIII.)

(Figure 9.11). Thus, all difficulties associated with meshing a given geometry are restricted to a lower-order manifold that constitutes the wetted surface of the geometry (Aftosmis *et al.* 1998).

In addition to the factors described in this section that influence the ability to access and visualize the data for immersive building simulation, data access and manipulation depends on an interaction mechanism that requires intuitive interface. This mechanism takes advantage of both special immersion and command functionality. The interaction mechanism will be described in the next section and issues related to specialty interface will be discussed briefly in the example case at the end of the chapter.

9.2.2 Interaction

For a successful immersive building simulation, issues of registration, display and latency have to be resolved. These are typically issues related to hardware, user and software interactions.

9.2.2.1   Registration

As mentioned earlier, registration requirements and needs are different for various immersive environments. In completely immersed environments, our eyes do not notice slight errors since we are not "in" the real scene; but in augmented environments, slight errors are noticed instantly due to the known phenomenon of our brain— "visual capture" (Welch and Robert 1978), where visual information overrides all other senses. In other words, the registration error of a Virtual Environment results in visual–kinesthetic and visual–proprioceptive conflicts rather than a visual–visual conflict as seen in the case of an AR system (Pausch *et al.* 1992). The sensitivity of human eyes to detect registration errors is extremely high (registration errors range from 0.1 to 1.8 mm for position and 0.05° to 0.5° for orientation) and this poses a great challenge in creating augmented environments without registration error. In AR, the range of space is limited by the sensor technology used. In general, registration errors are mainly due to sensors tracking, display configuration and viewing parameters. These issues will be disussed in the following sections.

SENSORS TRACKING

Sensors track the movement of an object or viewer in terms of position and orientation. Most commercial sensors have the capability to track six degrees of freedom (DOF) at any given time-interval. These sensors use different technologies such as mechanical, electromagnetic, ultrasonic, inertial, and optical. Each of these sensor technologies has limitations. Mechanical sensors are bounded by the connected device such as the BOOM. Electromagnetic sensors are prone to distortion due to metal present in the environment and propagate a high degree of error that changes with the distance between the sensors and the magnetic transmitter; ultrasonic sensors suffer from noise and temperature; the inertial sensors drift with time and cannot determine position (Figure 9.12). Although the optical sensors are comparatively

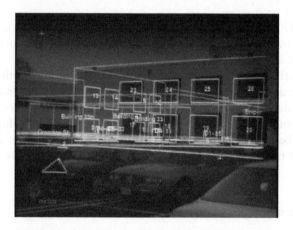

Figure 9.12 Outdoor AR-battlefield augmented reality system (inertial GPS technology) (Courtesy: The Advanced Information Technology Branch of Information Technology Division at the Naval Research Laboratory, US Navy). (See Plate XIV.)

Figure 9.13 Motion-capture of an actor performing in the liveActor (Optical technology) (Courtesy: Salim Zayat, University of Pennsylvania). (See Plate XV.)

Figure 9.14 Sensors adorn the actor's body (Courtesy: Kevin Monko, Thinkframe). (See Plate XVI.)

better than the others, they suffer from distortion and they are sensitive to object blocking. These sensors are typically used for motion-capture applications that track the movement of the body within a three-dimensional space (Figures 9.13 and 9.14).

Table 9.3 provides a sample comparison of different sensor types and their associated static accuracy position and orientation errors.

DISPLAY

Several display mechanisms exist. For fully and partial immersive environments, the most commonly used displays are HMD, BOOM and CAVE.

Head Mounted Display (HMD) is a complex physical system that integrates the combiners, optics and monitors into one (Figure 9.15). HMDs typically use either

Table 9.3 Sensor technologies and associated errors

Sensor technology	Description	Static accuracy position (RMS)	Static accuracy orientation (RMS)	Verified range
Magnetic	Flock of Birds (Ascension Technology Inc.)	1.8 mm	0.5°	20.3–76.2 cm
Ultrasonic	IS-600 Mark 2 PLUS (InterSense Inc.)	1.5 mm	0.05°	Not specified
Inertial	MT9 (Xsens Technologies B.V)	—	<1°	Not Specified
Optical	Laser Bird (Ascension Technology Inc.)	0.1 mm	0.05°	Static accuracy @ 1 m

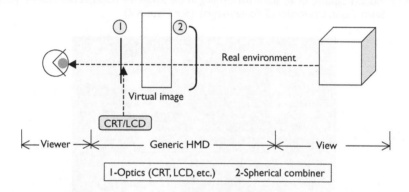

Figure 9.15 Generic HMD.

Liquid Crystal Display (LCD) or Cathode Ray Tubes (CRT). Any mechanical misalignment of the optical lenses or monitors will develop registration error. Stereographic projections with HMD need accurate positioning of the optics and depend on inter-pupillary distance of the eyes.

HMD are available in different resolutions, contrast rations, field of view, etc. High resolution HMD superimposes good non-pixellated images on the real scene allowing better visual perception of complex images. HMD can be used for both virtual and augmented environments. This is a function of the LCD or CRT of being able to allow the user to see through the optics.

In the Binocular Omni-Orientation Monitors (BOOMs), screens and optical system are housed in a box. This is attached to a multi-linked arm that contains six joints that enables the user to move the boom within a sphere of approximately a six-feet radius and balanced by a counterweight. Opto-mechanical sensors are located in the arm joints that enable the position and orientation of the BOOM to be computed. The user looks into the box through two holes, sees the virtual world, and can guide the box effortlessly to any position within the operational volume of the device. High

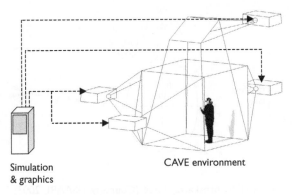

Simulation
& graphics

CAVE environment

Figure 9.16 A diagram of the CAVE environment.

quality CRT screens, wide field of view optics, superior tracking and reduced intrusiveness makes this device suitable for many applications where precision is needed.

The CAVE on the other hand, is a projection-based totally immersed environment that surrounds the viewer with four screens. The screens are arranged in a cube made up of three projection screens for walls and a projection screen for the floor (Figure 9.16). The projectors and the mirrors for the sidewalls are located behind each wall. The projector for the floor is suspended from the ceiling of the CAVE, which points to a mirror that reflects the images onto the floor. A viewer wears Stereographics' CrystalEyes liquid crystal stereo shutter glasses and a 6-DOF head-tracking device (the CAVE supports several types of tracking systems). As the viewer moves inside the CAVE, the correct stereoscopic perspective projections are calculated for each wall. The stereo emitters are placed around the edges of the CAVE. They are the devices that synchronize the stereo glasses to the screen update rate of 120 or 96 Hz. A wand (a 3D mouse) with buttons is the interactive input device. The primary wand has three buttons and a pressure-sensitive joystick. It is connected to the CAVE through a PC, which is attached to the supercomputer serial ports. A server program on the PC reads data from the buttons and joystick and passes them to the supercomputer.

The standard CAVE is a ten-foot cube. The origin of the coordinate system (0, 0, 0) for the CAVE is normally located at the center of the floor, that is five feet away from any wall. This means that the programmer has from +5 to −5 feet horizontally and from 0 to 10 feet vertically to define objects inside the CAVE. All the walls of the CAVE share the same reference coordinate system.

VIEW AND THE VIEWER

Due to eye anatomy, each user in the immersive environment can perceive the view slightly differently. The viewing parameters contribute an important component to the registration problem (Figures 9.17 and 9.18). These parameters include: the center of projection and viewport dimensions, offset between the location of the head tracker and the viewer's eyes and field of view (FOV).

Transformation matrices are used to represent the view registration. These matrices need to be aligned with the tracker system employed. For example, a magnetic

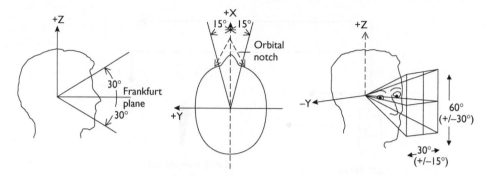

Figure 9.17 Vector limits of HMD breakaway force (Courtesy: USAARL 2003).

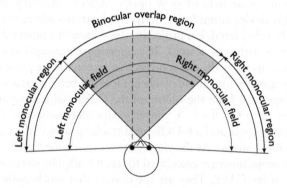

Figure 9.18 Human vision system's binocular FOV (Courtesy: USAARL 2003).

tracker such as Ascension-Tech's MotionStar Tracker performs with a set of axes, in which z-axis is "up". In Sun Java platform, the z-axis is in the direction of the depth of space (Figure 9.19). Incorrect viewing transformations pose registration problems to the immersive environment. The view might be perceived to be registered from a particular position in the working space and can be distorted from a different position. Such distortion can be eliminated by calculating the viewing transformation, both translation and orientation, from the eye position. In such cases, the view transformation will position the eye as the origin of rotation for the matrices to be deployed. Any difference in FOV of the user and the generated image will introduce registration errors. Moreover, the sequence of orienting the view platform is critical for proper viewing.

9.2.2.2 Latency

Latency is defined as the time difference between the sensor tracking the position and the posting of the respective image on the HMD. Human eyes are enormously fast

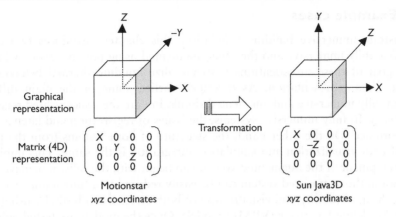

Figure 9.19 Transformation from hardware data to programming data.

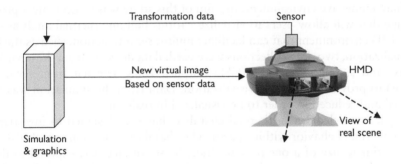

Figure 9.20 Latency due to system lag.

and process the new view immediately while the computer must generate images with new sets of transformation matrices from the sensors (Figure 9.20). Most of the latency issues are tied to the computer system delay (Holloway 1995). Moreover, the development platform, C++ or Java, is also vital to the processing of such complex programs (Marner 2002).

Predicting the future viewpoint (and thereby the viewing transformation) could resolve latency issues. One such predicting algorithm is the "Kalman filter", which implements a predictor–corrector type estimator that minimizes the estimated error-covariance when some presumed conditions are met. The use of linear accelerometers and angular rate gyroscopes to sense the rate of motion with the aid of Kalman filter (Kalman 1960) enables accurate immersive environments such as AR (Chai *et al.* 1999). Single-Constraint-at-a-Time (SCAAT) tracking integrated with Kalman filtering has shown improved accuracy and estimates the pose of HiBall tracker in real-time (Welch *et al.* 1999).

9.3   Example cases

To illustrate immersive building simulation, this chapter introduces two example cases, one fully immersive and the other augmented using post-processed CFD data.

The goal of these investigations is to visualize building thermal behavior using accurate data representation. As described earlier, some of the main differences between fully immersive and augmented simulation are the issues related registration and latency. In fully immersive systems, the issues of registration and latency are not of importance. On the other hand, the structure of both systems from the point of view of data visualization and simulation integration is the same. This implies that the development of the augmented system can begin with the fully immersive system. In addition, the augmented system can be easily reduced to a fully immersive system. Testing the applicability and visualization of both systems can begin by using Virtual Reality Modeling Language (VRML) models. Once the models are tested, additional functionality can be added and translated into the immersive or augmented hardware.

9.3.1   Fully immersive CFD visualization

For a fully immersive environment, the aim of this study was to generate a prototype technique that will allow users to visualize various building thermal analysis data in a virtual 3D environment that can facilitate multi-user interaction, such as the CAVE. For visualization, two data collections were used: data detected from the environment (sensors) and simulation results (CFD output). Data detected from sensors was enhanced to provide better data visualization. Data from the simulation was further processed and reduced in order to be visualized in real-time.

The space modeled was a thermal chamber that was designed to investigate the dynamic thermal behavior within spaces. The chamber dimensions are $8' \times 8' \times 8'$— the approximate size of a one-person office. Its south face was exposed to the outside. The other surfaces are under typical indoor conditions (Figures 9.21 and 9.22).

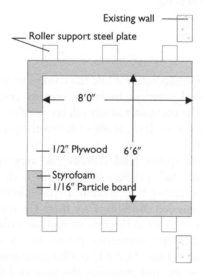

Figure 9.21 Test room—plan.

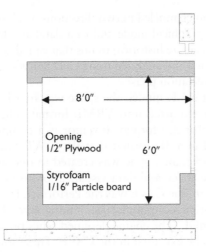

Figure 9.22 Test room—section.

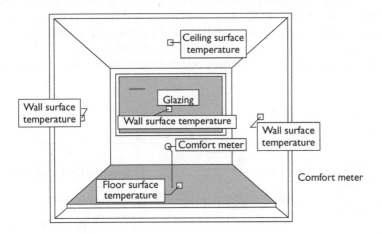

Figure 9.23 Test room—sensor locations.

The room was designed to provide flexibility for supply air diffuser location, amount and location of thermal mass, interior finish material as well as glass type. The room is equipped with a ducted fan coil air condition unit with a hot water booster heating coil. Electric water heaters and chillers are connected with heating and cooling coils respectively. All the mechanical system components can be controlled either manually or through the Building Modular Controller.

Five thermocouples were used to measure the room surface temperatures with 0.5 °F accuracy. These sensors were placed on the walls and window of the test room (Figure 9.23). One thermal comfort transducer—connected to a thermal comfort meter—was placed in the middle of the room. The information from these sensors was collected in real-time and channeled in two directions: back to the interface for the user to view the results in a graphical mode and to a database. Information from the database was then used to generate historical trends that can also be displayed. This thermal chamber was also modeled and analyzed using CFD Software. The data output was then customized for the visualization phase.

In order to construct a VR model, the space was first built using 3D Studio Max and then the data was exported into VRML format. The VRML model was then transformed into Inventor 2.0 format. Inventor is a computer program that allows objects to be displayed in a 3D format that the CAVE can display. Once the model became CAVE compatible, an engine was created to operate the CAVE to allow the interaction between the space and data visualization to be displayed. The computer language used to program the CAVE was the Performer. It is a computer graphics language that allows real-time communication among several programming languages, such as C++ and OpenGL. The engine designed relies on several files that behave as the source files for it to be executed in the CAVE. These source files are connected to the CAVE libraries. Hence, they provide the specifications for some of the basic configurations of the CAVE such as the speed of navigation, the buttons functions, and the interaction with the space.

When executing the engine, two different calls occur at the same time. The thermal data plotting (from sensors and from simulation) and the model of the space, which is functioning as the background of the thermal analysis. These data are then mapped onto the space as one object using the performer functions in the CAVE. This allows the user to move around both the data and the room at the same time, Figure 9.24.

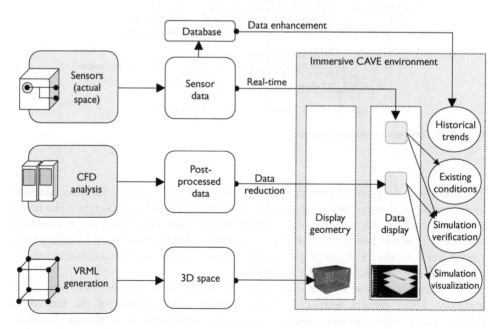

Figure 9.24 Data computing procedure.

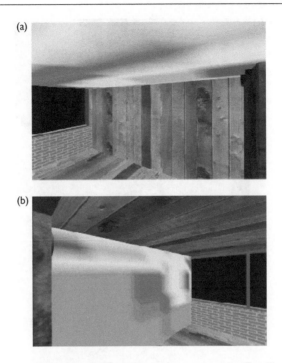

Figure 9.25 CAVE display of CFD thermal data as seen by participants. (See Plate XVII.)

Sensors channel data into the model for simulation verification as well as for storage into a database that can be called on to visualize historical trends. For data visualization, three different 2D meshes were created (Malkawi and Primikiri 2002), one in each direction (*xy*, *yz*, *xz*). The main technique used for the meshes was tristrips. This technique allowed some flexibility in terms of the mesh control and thus enabled the possibility of manipulating the nodes of the mesh more easily. Once the temperature data from the CFD analysis or database is sent to the Performer, it is stored in a 3D array and then assigned to each of the nodes of the mesh. According to the temperature value, a color range was assigned and displayed (Figure 9.25 (a) and (b)).

One or more users were able to navigate through the space, visualize its resulting thermal conditions, real and simulated, and view historical trends. In addition, users could change some of the space parameters such as window size or materials and visualize the resulting thermal conditions. These conditions are displayed by choosing the desired mesh to be deployed. With the click of the buttons on the wand, the user can decide exactly which slice to show, visualize this information and share it with other users. Simulation information regarding that condition is extracted and displayed in a 3D format (Figures 9.26 (a) and (b)).

9.3.2 *Augmented simulation and visualization*

This study extends on a history for attempting to build a system that allows the human eye to detect the impact of environmental conditions or perceiving the different

(a)

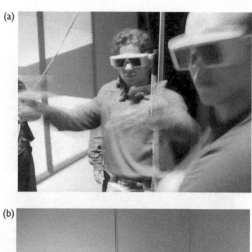

(b)

Figures 9.26 Participants inside the CAVE environment. (See Plate XVIII.)

environmental factors in the three-dimensional geometry of the work place (Malkawi and Choudhary 1999; Malkawi and Primikiri 2002). As mentioned earlier, Augmented Reality is an emerging new Human–Computer Interface paradigm that has the potential to achieve such a goal.

To develop an augmented system, one can test the data visualization capability and then the issues of integration between simulation and visualization in real-time can be investigated. Data mapping techniques similar to the ones used in fully immersive environments can be used for the augmented systems. To account for the real-building overlay, no geometrical representation will be required. Only data visualization and interaction paradigms need to be developed. Interaction methods and behavior can be tested using VRML and then translated into the hardware.

To test the simulation interaction with visualization, one can develop a VRML model, test its methods and then translate it into AR modules. One such study is a system that was developed to allow users to remedy problems on site by providing the ability to try various combinations of building components and test their impact within an interior space (Malkawi and Choudhary 1999). This system allows users to manipulate building envelope parameters and materials, invoke a simulation model and display the heat transfer information and comfort condition using the VRML. The objective is to develop techniques for the visualization of thermal simulations in a highly interactive manner. In addition, the computational representation for rapid determination of the impact of changes in building parameters was the primary concern.

The model contains a Java module and an interface module. The Java module contains the simulation engine and functions that manipulate the interface and has been implemented as an object-oriented module. Two numerical models are included within the simulation. The first model determines the discomfort index as an array of preset nodes within the room. The second model computes the peak heating or cooling load for a representative hour of the month.

The computational model emphasizes upon the role of building enclosure in modifying the climatic variables such as air temperature, radiation, and humidity to suit comfort conditions. As a result, the computational model allows the user to vary the physical properties of the enclosing surfaces of a space, and configurations of opaque and transparent components for a given climatic context. An array of preselected nodes within the room interior was embedded into the computational model to determine comfort levels at different locations within the room. The output is presented to the user in the form of a three-dimensional color graph overlaid on the floor area of the room.

Heating and cooling loads are also computed and displayed (Malkawi and Choudhary 1999). Results are displayed in the form of graphs overlaid on the wall surface for the peak hour of each month. Once the user commits to a certain choice of building parameters, the simulation is performed and output is updated in real-time in the virtual environment. Several functionalities were included such as a check of simulation input accuracy and compliance.

Based on this work two studies took place. The first is a direct translation of the VRML model to an augmented system using an optical see-through LCD head-mounted display. The second study mapped CFD output with the see-through CRT head-mount display.

For the first study, the AR system that was developed allows the user to use the same interface developed for the VRML model to manipulate the building parameters and display results on the actual room. The optical combiners in front of the eye allow light in from the real world and they also reflect light from monitors displaying graphical images. The result is a combination of the real world and a virtual world drawn by the monitors. The thermal data generated had to be accurately registered with the real world in all dimensions. The correct registration had to be maintained while the user moves within the real environment. The simulations and databases developed in Java were placed in a client–server environment. The output of the simulation was superimposed on the real room. The user can change the parameters described earlier to control the simulation. The AR system allows the user to perform several tasks. It allows running a simplified simulation on the fly, changing the parameters of the physical environment such as room material and sending the output of the simulation to the HMD for the user to visualize the information on top of the actual scene.

To allow higher interaction level and to work with more complicated simulations, the second study was developed to integrate CFD output with a see-through CRT head-mount display (Figures 9.27 and 9.28).

The goal is to integrate a real-time CFD engine with sensor connection to boundary conditions to allow on the fly data visualization of real environments. This implies the use of dynamic immersive simulation which introduces many challenges as described in

Figure 9.27 Viewer with HMD. (See Plate XIX.)

Figure 9.28 Registration of room in real-space. (See Plate XX.)

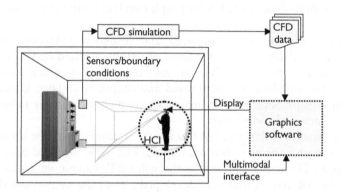

Figure 9.29 Integration of multimodal interface with AR environment.

earlier sections. Some of these challenges are being researched such as integrating sensors and boundary conditions for automatic updating and calibration. This allows the user to view sensor information such as temperature and airflow in real-time and compare it with the simulated results. In addition, interactions with the CFD output require a new intuitive method of interfacing that is specific for the application (Figure 9.29). This involves the integration of a multimodal HCI by transforming human perceptual cues such as hand movements (haptic), speech recognition, etc. to a set of functions for effective data visualization within the AR environment.

The CFD output was generated as a VRML file, scaled and oriented such that it is accurately mapped with the real scene, and posed into the HMD. The VRML importer communicates with the CFD engine through loading classes that receive input from the user about the orientation and position of the data requested and passes this information on to the command generator. The generator creates a set of command-line directions and calls the CFD engine to execute the file. The results are saved in a specified location in a VRML format. The data is then imported into Java3D using a plug-in package to convert the data from VRML to native Java3D

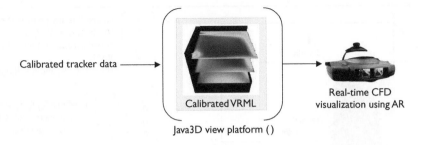

Calibrated tracker data ——————→

Calibrated VRML

Java3D view platform ()

Real-time CFD
visualization using AR

Figure 9.30 VRML scaling and Java3D.

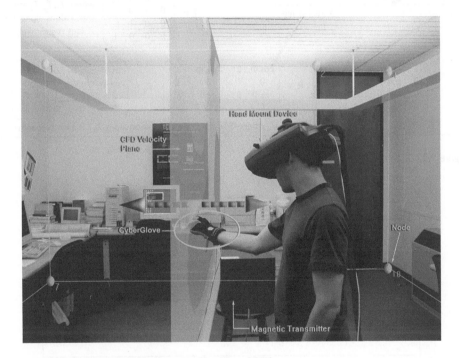

Figure 9.31 Interfacing with CFD datasets. (See Plate XXI.)

format (Figure 9.30). The result is a system that allows the user to visualize and manipulate CFD data on actual environment (Figure 9.31).

There are two processes that take place before the model is sent to the HMD. The first is model calibration and the second is tracker data smoothing. For calibration, a 4D matrix representation is used and scaling (s), rotation (r) and translation (l) transformations are applied to the VRML (v_n) in order to fit the real space (Figure 9.32).

For the smoothing process of the 6 DOF Tracker data (R), Gaussian filter is used. Gaussian filter acts as a "point-spread" function, and reduces the noise by convolution. The Gaussian outputs a "weighted average" of each 6 DOF data's neighborhood, thus enabling smoothing of 6 DOF real-time data (Figure 9.33).

There are a number of limitations to the full functionality of the system. The data from the tracker is being streamed directly into the engine, which produced a close to

$$P_i = \begin{pmatrix} X_i \\ Y_i \\ Z_i \\ V \end{pmatrix}$$

where P_i is any point in the VRML, and X_i, Y_i, Z_i are coordinates and the Calibrated VRML
$(V) = T\,R\,S\,V_n$ and $R = R_y R_x R_z$

$$S = \begin{pmatrix} S_x & 0 & 0 & 0 \\ 0 & S_y & 0 & 0 \\ 0 & 0 & S_z & 0 \\ 0 & 0 & 0 & 1 \end{pmatrix}$$

where S_x, Y_y, S_z are scale factors of X,Y,Z axes respectively.

$$L = \begin{pmatrix} 1 & 0 & 0 & L_x \\ 0 & 1 & 0 & L_y \\ 0 & 0 & 1 & L_z \\ 0 & 0 & 0 & 1 \end{pmatrix}$$

where T_x, T_y, T_z are translations along X,Y, Z axes respectively.

$$R_x = \begin{pmatrix} 1 & 0 & 0 & 0 \\ 0 & \cos\Phi & \sin\Phi & 0 \\ 0 & -\sin\Phi & \cos\Phi & 0 \\ 0 & 0 & 0 & 1 \end{pmatrix} \quad R_y = \begin{pmatrix} \cos\Phi & 0 & -\sin\Phi & 0 \\ 0 & 1 & 0 & 0 \\ \sin\Phi & 0 & \cos\Phi & 0 \\ 0 & 0 & 0 & 1 \end{pmatrix} \quad R_z = \begin{pmatrix} \cos\Phi & \sin\Phi & 0 & 0 \\ -\sin\Phi & \cos\Phi & 0 & 0 \\ 0 & 0 & 1 & 0 \\ 0 & 0 & 0 & 1 \end{pmatrix}$$

where Φ is the angle of rotation along X,Y, Z axes respectively.

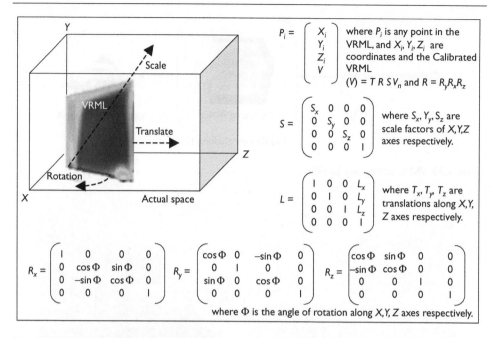

Figure 9.32 VRML calibration.

Tracker data $(R_t) = \{t_1(x_1\, y_1\, z_1\, a_1\, b_1\, c_1)\}_1^{\alpha}$ where x_1, y_1, z_1 are sensor positions, and $a_1\ b_1\ c_1$ are sensor orientations along X, Y, Z, at time $t = 1$.

Gaussian 1D Filter $G(x) = \left(\frac{1}{\sqrt{2\pi}\,\sigma}\ e^{-x^2/2\sigma^2} \right)$

The convolution is performed by sliding the kernel $G(x)$ over R, in real-time. Thus the

Calibrated Tracker data $(R) = \sum_{t=1}^{\alpha} R(t-x) \cdot G(x)$

Figure 9.33 Tracker data smoothing.

real-time refresh rate with small lag time. This calls for other methods of data delivery to produce more acceptable refresh rates for the system. The second limitation is the data display. The system supports only iso-surface cuts along the x, y and z axes at specified distances. This can be overcome by expanding the system beyond reliance on the VRML output to work with the raw data that will allow the user to specify more complex data visualization. The system will only function in a predetermined location. This also calls for developing a method that allows it to be independent of locale and allow visualization data to be gathered on the fly.

Currently, functions that allow intuitive and robust system for haptic interaction with 3D CFD environments are being developed. This includes interactive speech and gesture recognition that will allow the user to interact with the raw data of the CFD simulation. In addition, statistical filters are being deployed to reduce noise and achieve better smoothing to allow a higher degree of registration and real-time data visualization.

9.4 Conclusions

This chapter discussed the background of immersive building simulation and its challenges. It introduced the elements that are crucial in the development of a simulation environment that is still in its infancy stage. The few studies that were conducted in this area illustrate the potential application of such research. Major challenges still exist in regard to the software and hardware, as well as the knowledge required to bring this to the building simulation field. These challenges are

- *Hardware cost*: The cost of hardware required to develop immersive building simulation environments is still high and few vendors provide this specialty hardware.
- *Expertise*: Different expertise is required to construct such environments. This requires collaborative effort between a variety of researchers.
- *Interaction methods*: Human–computer interface for immersion environments and buildings is not available. Interaction methods need to be developed in order to take advantage of the immersion component.
- *Sensor technology*: Actual building conditions through sensor data and simulated behavior through simulation models can inform each other, but integration of sensors and simulation needs to be further developed.
- *Data visualization*: Developing data reduction and enhancement techniques that can be used for real-time dynamic simulation is needed.
- *Software*: Software that supports building simulation objects and methods is not available. Using existing methods is a challenging task even for simple operations. For example, translating information from the CFD simulation to the CAVE required interface mapping and the CAVE language is also restrictive regarding the primitives it uses. Only basic shapes can be used, such as cubes or spheres. Thus it makes it difficult to map information with complex shapes to the output of the CFD simulations.
- *Incorporation with building product models*: To allow robust and fast development standards for data communications need to be developed.

Despite these challenges, immersive environments provide opportunities that are not available using current simulation models and interactions. These environments prove the potential for their use in collaboration. For example, using the CAVE allowed multi-users to virtually navigate through a 3D environment and share information regarding its performance.

Immersive simulation extends on the 3D performance information visualization of buildings and permits the user to interact with the real environment and visualize the reaction of this interaction. It presents a new way of interfacing with the built environment and controlling its behavior in real-time. This will become evident as more techniques such as optimization and knowledge-based systems, etc. give these environments more power as users navigate through them and interact with their elements.

References

Aftosmis, M.J, Berger, M.J., and Melton, J.E. (1998). "Robust and efficient Cartesian mesh generation for component-based geometry." *AIAA Journal*, Vol. 36, pp. 952–960.

Azuma, T.R. (1993) "Tracking requirements for augmented reality." *Communications of the ACM*, Vol. 36, No. 7, pp. 150–151.

Azuma, T.R. and Bishop, G. (1995). "A frequency-domain analysis of head-motion prediction." In *Proceedings of SIGGRAPH '95, Computer Graphics, Annual Conference Series*, Los Angeles, CA, 6–11, August, pp. 401–408.

Azuma, T.R. (1997). "A survey of augmented reality." *Presence: Teleoperators and Virtual Environments*, Vol. 6, No. 4, pp. 355–385.

Azuma, T.R., Baillot, Y., Behringer, R., Feiner, S., Julier, S., and MacIntyre, B. (2001). "Recent advances in augmented reality." *IEEE Computer Graphics and Applications*, Vol. 21, No. 6, pp. 34–47.

Berger, M.O., Wrobel-Dautcourt, B., Petitjean, S., and Simon, G. (1999). "Mixing synthetic and video images of an outdoor urban environment." *Machine Vision and Applications*, Vol. 11, No. 3, pp. 145–159.

Chai, L., Nguyen, K., Hoff, H., and Vincent, T. (1999). "An adaptive estimator for registration in augmented reality." In *Second IEEE and ACM International Workshop on Augmented Reality (IWAR)*, San Francisco, CA, October 20–21.

Chen, H.L. and Ho, J.S. (1994). "A comparison study of design sensitivity analysis by using commercial finite element programs." *Finite Elements in Analysis and Design*, Vol. 15, pp. 189–200.

Chen, J., Lobo, N., Hughels, C., and Moshell, J. (1997). "Real-time fluid simulation in a dynamic virtual environment." *IEEE Computer Graphics and Applications*, pp. 52–61.

De Leeuw, W.C. and van Wijk, J.J. (1993). "A probe for local flow field visualization." In *Proceedings of Visualization 1993*, IEEE Computer Society Press, Los Alamitos, CA, pp. 39–45.

Fiener, Webster, Kruger, MacIntyre, and Keller. (1995). "Architectural anatomy." *Presence* 4(3), pp. 318–325.

Frost, P. and Warren, P. (2000). "Virtual reality used in a collaborative architectural design process." In *Proceedings of Information Visualization*, IEEE International Conference 2000, pp. 568–573.

Fuhrmann, A. and Loffelmann, H. (1998). "Collaborative visualization in augmented reality." *IEEE Computer Graphics and Visualization*, pp. 54–59.

Giallorenzo, V. and Banerjee, P. (1999). "Airborne contamination analysis for specialized manufacturing rooms using computational fluid dynamics and virtual reality." *Journal of Material Processing and Manufacturing Science*, Vol. 7, pp. 343–358.

Geiben, M. and Rumpf, M. (1992). "Visualization of finite elements and tools for numerical analysis." In F.H. Post and A.J.S. Hin (eds) *Advances in Scientific Visualization*, Springer, pp. 1–21.

Haber, R.B. and McNabb, D.A. (1990). "Visualization idioms: a conceptual model for scientific visualization systems." In G.M. Nielson, B. Shriver, and L.J. Rosenblum (eds) *Visualization in Scientific Computing*. IEEE Computer Society Press, Los Alamitos, CA, pp. 74–92.

Hesselink, L., Post, F.H., and Wijk, J.J.van. (1994). "Research issues in vector and tensor field visualization." In Visualization Report. *IEEE Computer Graphics and Applications*, pp. 76–79.

Hillier, Frederick, S. and Liberman, J. (1995). *Introduction to Mathematical Programming*. McGraw-Hill Inc., New York.

Holloway, R. (1995). "Registration Errors in Augmented Reality." PhD dissertation. UNC Chapel Hill, Department of Computer Science Technical Report TR95-016.

Impelluso, T. (1996). "Physically based virtual reality in a distributed environment." *Computer Graphics*, Vol. 30, No. 4, pp. 60–61.

Jacob, M. (2002). Evaluating java for game development. M.Sc. thesis. Department of Computer Science, University of Copenhagen, Denmark.

Kalman, R.E. (1960). "A new approach to lineaer filtering and prediction problems." *Transaction of the ASME—Journal of Basic Engineering*, pp. 35–45.

Kenwright, D. (2000). "Interactive computer graphics." *Aerospace America*, AIAA, p. 78.

Klinker G., Stricker, D., Reiners, D., Koch, R., and Van-Gool, L. (1998). "The use of reality models in augmented reality applications. 3D structures from multiple images of large scale environments." In *Proceedings of the European Workshop SMILE 1998*, Springer-Verlag, Berlin, Germany, viii +346, pp. 275–289.

Krueger, M. (1985). "VIDEOPLACE: a report from the artificial reality laboratory." *Leonardo*, Vol. 18, No. 3, pp. 145–151.

Krueger, M. (1991). *Artificial Reality II*. Addison and Wesley, Reading, MA.

Linden, E., Hellstrom, J., Cehlin, M., and Sandberg, M. (2001). "Virtual reality presentation of temperature measurements on a diffuser for displacement ventilation." In *Proceedings of the 4th International Conference on Indoor Air Quality and Energy Conservation in Buildings*, Changsha, Hunan, V.II., pp. 849–856.

Malkawi, A. and Choudhary, R. (1999). "Visualizing the sensed environment in the real world." *Journal of Human-Environment Systems*, Vol. 3, No. 1, pp. 61–69.

Malkawi, A. and Primikiri, E. (2002). "Visualizing building performance in a multi-user virtual environment." In *Proceedings of the ACSA International Conference*, Havana, Cuba.

McCormick, B., DeFanti, T.A., and Brown, M.D. (eds) (1987). "Visualization in scientific computing." *Computer Graphics*, Vol. 21(6), pp. 1–14.

Milgram, P. and Kishino, P. (1994). "A taxonomy of mixed reality visual displays." *IEICE Transactions on Information Systems*, Vol. E77-D(12), pp. 1321–1329.

Milgram, P., Takemura, H. *et al.* (1994). "Augmented reality: a class of displays on the reality-virtuality continuum." *SPIE Proc.: Telemanipulator and Telepresence Technologies*, 2351m, pp. 282–292.

Neil, M.J. (1996). "Architectural virtual reality applications." *Computer Graphics*, Vol. 30, No. 4, pp. 53–54.

Oliver, (2003). Figure of hybrid mesh. Available online at http://www.vug.uni-duisburg.de/~oliver/welcome.html (accessed on June 15, 2003).

Pagendarm, H.G. (1993). "Scientific visualization in Computational Fluid Dynamics." Reprinted from J.J. Connors, S. Hernandez, T.K.S. Murthy, and H. Power (eds) *Visualization and Intelligent Design in Engineering and Architecture*, Computational Mechanics Publications, Elsevier Science Publishers, Essex, United Kingdom.

Pausch, Randy, Crea, T. and Conway, M. (1992). "A literature survey for virtual environments: military flight simulator visual systems and simulator sickness." *Presence: Teleoperators and Virtual Environments*, Vol. 1, No. 3, pp. 344–363.

Pilgrim, M., Bouchlaghem, D., Loveday, D., and Holmes, M. (2001). "A mixed reality system for building form and data representation." *IEEE*, pp. 369–375.

Post, F.H. and van Wijk, J.J. (1994). "Visual representation of vector fields: recent developments and research directions." In L. Rosenblum, R.A. Earnshaw, J. Encarnacao, H. Hagen, A. Kaufman, S. Klimenko, G. Nielson, F.H. Post, and D. Thalman, (eds) Chapter 23 in *Scientific Visualization—Advances and Challenges*, IEEE Computer Society Press.

Rangaraju, N. and Tek, M. (2001). "Framework for immersive visualization of building analysis data." *IEEE*, pp. 37–42.

Reed, D., Giles, R., and Catlett, C. (1997). "Distributed data and immersive collaboration." *Communications of the ACM*, Vol. 40, No. 11, pp. 39–48.

Retik, A., Mair, G., Fryer, R., and McGregor, D. (1998). "Integrating Virtual Reality and Telepresence to Remotely Monitor Construction Sites: a ViRTUE Project." *Artificial Intelligence in Structural Engineering, Information Technology for Design, Collaboration, Maintenance and Monitoring*, Springer-Verlag, Berlin, Germany, pp. 459.

Shahnawaz, V.J., Vance, J.M., and Kutti, S.V. (1999). "Visualization of post processed CFD data in a virtual environment." In *Proceedings of DETC99*, ASME Design Engineering Technical Conference, Las Vegas, NV, pp. 1–7.

Saltelli, A., Chan, K., and Scott, E.M. (eds) (2000). *Sensitivity Analysis*. Wiley Publications.

Schroeder, R. (1993). "Virtual reality in the real world: history, applications and projections." *Futures*, p. 963.

Stalker, R. and Smith, I. (1998). "Augmented reality applications to structural monitoring." *Artificial Intelligence in Structural Engineering, Information Technology for Design, Collaboration, Maintenance and Monitoring*, Springer-Verlag, Berlin, Germany, p. 479.

Sun, D. (2003). "Multiblock." Available online at http://thermsa.eng.sunysb.edu/~dsun/ multiblock.html (accessed on June 5, 2003).

Sutherland, I.E. (1965). "The ultimate display." In *Proceedings of the 1965 IFIP Congress*, Vol. 2, pp. 506–508.

USAARL (2003). Figure on "Vector limits of HMD breakaway force" from Chapter 7. Biodynamics; Figure on "Human vision system's binocular FOV" from Chapter 5. Optical Performance; Part Two—Design Issues Optical Performance, *HMDs: Design Issues for Rotary-Wing Aircraft*. Available online at http://www.usaarl.mil/hmdbook/cp_005.htm (accessed on June 4, 2003).

van Dam, Forsberg, A., Laidlaw, A., LaViola, J.D., and Simpson, R.M. (2000). "Immersive VR for scientific visualization: a progress report." *IEEE Computer Graphics and Applications*, Vol. 20, Issue 6, pp. 26–52.

Walsum, Van. T. (1995). Selective visualization on curvilinear grids. PhD thesis. Delft University of Technology.

Walsum Van. T., Post, F.H., and Silver, D. (1996). "Feature extraction and iconic visualization." *IEEE Transactions on Visualization and Computer Graphics*, Vol. 2, No. 2, pp. 111–119.

Welch, R. (1978). *Perceptual Modifications: Adapting to Altered Sensory Environments*. Academic Press.

Welch, G., Bishop, G., Vicci, L., Brumback, S., Keller, K., and Colucci, D'nardo (1999). "The HiBall Tracker: High-performance wide-area tracking for virtual and augmented environments." *Symposium on Virtual Reality Software and Technology*, University College, London.

Epilogue

Godfried Augenbroe and Ali M. Malkawi

The nine chapters in this book have provided an account of advanced research topics in building simulation. The overview covers many of the pertinent issues but is by no means intended to be exhaustive.

The authors have made a set of important observations about where the field will be heading. They have stressed the many opportunities for further maturation of the building simulation discipline. Some authors give priority to tool functionalities and usability issues; others emphasize the professional deployment of tools and emphasize the need for functional integration of it in the design process. In spite of the progress in robustness and fidelity of the simulation toolset, there remain many targets that have not yet been achieved. Among the ones mentioned in the book are: support for rapid evaluation of alternative designs, better adaptation to decision-making processes, support of incremental design strategies, and improved quality assurance by validity constraints imposed by the application.

The need to apply increased rigor to the process of simulation has been stressed in various chapters and in different settings. The issue is strongly linked to quality assurance procedures and better management of the inherent uncertainties in the inputs and modeling assumptions in simulation.

The confidence in the coupling of multi-domain models is growing but it remains a necessary and attractive field of continued research, especially in the field of CFD. Other couplings, for example, with occupant models, mold growth, and contaminant spreading models, as well as with integral building automation systems and self-aware buildings deserve continued research efforts.

Work on interoperability has made great strides, but needs to be followed up by systems that harness interoperable tools in flexible systems that support a comprehensive dialogue between different actors in a building project team.

The shift toward new manifestations of simulation has been addressed in several chapters from different perspectives. Some emphasized the need for systems that support decision-makers, anticipating that embedded simulation will be key to support decision-making, and act as the core of model-based adaptable control systems. New methods to visualize simulation results and interact with real-time simulation are emerging. New environments will eventually allow a user while being immersed in the real environment, to interact with a running simulation of himself and his environment. It will become common place to interact with the virtual world around us, and interrogate a simulation model about the consequences of the proposed system intervention one is about to make. This is just one manifestation of "invisible" and ubiquitous simulation of which more examples have been given in the chapters.

The desire for shared software development and component sharing is still alive but not addressed directly in any of the chapters. New object-oriented environments are considered as the paradigm that could make it happen, especially in combination with web-hosted services, as these seem to hold a strong promise for "functional sharing" rather than code sharing.

Last but not least, it should be acknowledged that there is an educational agenda that needs to be pursued with vigor. All chapters reveal important challenges for the educational settings to teach building simulation in graduate and undergraduate programs. Cross-domain teaching linking education in building physics, human behavior and environment, architectural design, building and HVAC engineering, computer science, risk analysis, policy development, etc. is more important than ever.

The field is evolving and changing rapidly, and its advancement is influenced by the dynamics of how buildings are designed, conceived, and serviced.

Index